Personnel

AWS Committee on Definitions and Symbols

W. L. Green, Chairman	Ohio State University
R. L. Holdren, 1st Vice Chairman	Welding Consultants, Incorporated
A. J. Kathrens, 2nd Vice Chairman	Canadian Welding Bureau
C. B. Pollock, Secretary	American Welding Society
*J. T. Biskup***	Consultant
H. B. Cary	Hobart Brothers Company
*R. J. Christoffel**	Consultant
*G. B. Coates**	General Electric Company
R. B. Corbit	General Public Utilities Nuclear Corporation
C. K. Ford	Hobart School of Welding Technology
J. E. Greer	Moraine Valley Community College
M. J. Grycko, Jr.	Packer Engineering Associates
J. G. Guenther	Dean Lally LP
E. A. Harwart	Consultant
*M. J. Houle**	M. J. Houle, Incorporated
R. D. McGuire	National Board of Boiler and Pressure Vessel Inspectors
*G. E. Metzger***	Wright Patterson AFB
*D. H. Orts**	Armco, Incorporated
*E. J. Seel**	Consultant
L. J. Siy	U.S. Composites Corporation
J. J. Stanczak	Steel Detailers and Designers
J. J. Vagi	Engineering Consultant

Subcommittee on Definitions

R. L. Holdren, Chairman	Welding Consultants, Incorporated
J. E. Greer, Vice Chairman	Moraine Valley Community College
C. B. Pollock, Secretary	American Welding Society
*J. T. Biskup***	Consultant
H. B. Cary	Hobart Brothers Company
*A. B. Cedilote**	Industrial Testing Laboratory Services
*R. J. Christoffel**	Consultant
R. B. Corbit	General Public Utilities Nuclear Corporation
*A. T. Cullison**	AWS — Welding Journal
M. J. Grycko, Jr.	Packer Engineering
*M. J. Houle**	M. J. Houle, Incorporated
*D. L. Kuruzar**	Manufacturing Technology, Incorporated
*R. D. McGuire**	National Board of Boiler and Pressure Vessel Inspectors
*G. E. Metzger***	Wright Patterson AFB
*D. H. Orts**	Armco, Incorporated
*E. J. Seel**	Consultant
J. J. Vagi	Engineering Consultant

* Advisor
**Deceased

Dedication

This revision of AWS A3.0, Standard Welding Terms and Definitions, is respectfully dedicated to the memory of Dr. Guinn E. Metzger who served the A2 Committee as Chairman from 1983–1988 and the A2B Subcommittee as Chairman from 1979–1992. The devoted and zealous leadership he provided during those years advanced the concept and use of standard welding terms to unprecedented levels.

Foreword

(This Foreword is not a part of ANSI/AWS A3.0-94, *Standard Welding Terms and Definitions*, but is included for information purposes only).

The Committee on Definitions and Symbols was formed by the American Welding Society to establish standard terms and definitions to aid in the communication of welding information. This publication is the major product of work done by the Subcommittee on Definitions in support of that purpose.

The first AWS document containing welding definitions was prepared by the Committee of Definitions and Chart and approved by the Executive Committee as *Tentative Definitions of Welding Terms and Master Chart of Welding Processes*, on January 18, 1940. A revision was approved by the AWS Board of Directors on May 7, 1942.

The next revision, bearing the designation A3.0, was called *Standard Welding Terms and Their Definitions*. This revision, published in 1949, listed the terms alphabetically.

During the late 1950's, the Committee was reorganized as the AWS Committee on Definitions and Symbols, and after several years' work, produced A3.0-61, *AWS Definitions, Welding and Cutting*. Subsequent revisions were published in 1969, 1976, 1980, 1985, and 1989.

The present publication, A3.0-94, *Standard Welding Terms and Definitions*, defines 1253 terms, with 53 illustrations to support and clarify the definitions, as well as classification charts and corollary information for the welding processes.

The standard terms and definitions published here are those that should be used in the oral and written language of welding. Since this is intended to be a comprehensive compilation of welding terminology, nonstandard terms used in the welding industry are included. All terms are either standard or nonstandard; standard terms are identified by **boldface** and nonstandard terms are labeled as such.

Standard terms have been approved by the American Welding Society, whereas, nonstandard terms have not. It is recommended that standard terms be used in all welding literature, in particular, documents of a legal nature; for example, standards, contracts, laws, and regulations. Nonstandard terms are not recommended for any purpose.

To make this document most useful, the terms are arranged in the conventional dictionary letter-by-letter alphabetical sequence. It is the policy of the American Welding Society to use only generic terms and definitions in this publication. The numerous proprietary brand and trademark names commonly used to describe welding processes, equipment, and filler metals are not included.

To preserve an understanding of old documents and literature, welding terms believed to be no longer significant in the welding industry are included. Obsolete or seldom used processes are listed separately in Table 5.

The figures have been grouped together, rather than dispersed throughout the text. This grouping is desirable, since several figures illustrate more than one term.

Also, figures concerning related terms have been grouped to illustrate similarities, differences, and interrelationships.

Two classification arrangements are presented in this publication as charts of a hierarchy of welding processes.

The traditional Master Chart of Welding and Allied Processes places the main categories in the center, with the subcategories in boxes around the perimeter.

The Joining Method Chart and corollary classification charts are based exclusively on the physical state of materials at the joint during coalescence. This results in three major classifications of the welding processes; fusion welding for liquid/liquid, solid-state welding for solid/solid, and brazing and soldering for liquid/solid.

Some of the views and intentions of the Definitions Subcommittee are presented in the Definitions Subcommittee Manifesto, included in this document as Annex A.

To improve understanding of the terms and definitions published in A3.0, the Guide to A3.0 is included as Annex B.

Interested readers are encouraged to submit pertinent comments, including new or modified definitions to the Secretary, AWS Committee on Definitions and Symbols, American Welding Society, 550 N.W. LeJeune Road, Miami, FL 33126.

Table of Contents

Page No.

List of Figures

Figure **Page No.**

Page No.

List of Charts and Tables

Terms and Definitions

Note: Vertical lines preceding terms indicate where changes (additions, modifications, corrections, deletions) from the 1989 edition were made. A single line represents a minor or editorial change. Double lines represent a new term or a major change.

|| **Boldface** indicates standard terms, lightface indicates nonstandard terms. Terms for standard welding processes and for standard welding process variations are followed by their standard letter designations.

A

abrasion soldering. A soldering process variation during which the faying surface of the base metal is mechanically abraded.

abrasive blasting. A method of cleaning or surface roughening by a forcibly projected stream of abrasive particles.

absorptive lens. A filter lens designed to attenuate the effects of glare and reflected and stray light. See also **filter plate.**

accelerating potential, *electron beam welding and cutting.* The potential that imparts velocity to the electrons.

| **acceptable weld.** A weld that meets the applicable requirements.

| **acetylene feather.** The intense white, feathery-edged portion adjacent to the cone of a carburizing oxyacetylene flame. See Figure 40.

acid core solder. A solder wire or bar containing acid flux as a core.

activated rosin flux. A rosin base flux containing an additive that increases wetting by the solder.

active flux, *submerged arc welding.* A flux from which the amount of elements deposited in the weld metal is dependent on the welding conditions, primarily on the arc voltage. See also **neutral flux.**

actual throat. The shortest distance between the weld root and the face of a fillet weld. See Figure 25.

adaptive control, *adj.* pertaining to process control that automatically determines changes in process conditions and directs the equipment to take appropriate action. See also **automatic, manual, mechanized, robotic,** and **semiautomatic.**

adaptive control brazing. See **adaptive control welding.**

adaptive control soldering. See **adaptive control welding.**

adaptive control thermal cutting. See **adaptive control welding.**

adaptive control thermal spraying. See **adaptive control welding.**

adaptive control welding. Welding with a process control system that automatically determines changes in welding conditions and directs the equipment to take appropriate action. Variations of this term are **adaptive control brazing, adaptive control soldering, adaptive control thermal cutting,** and **adaptive control thermal spraying.** See Table 4. See also **automatic welding, manual welding, mechanized welding, robotic welding,** and **semiautomatic welding.**

air acetylene welding (AAW). An oxyfuel gas welding process that uses an air-acetylene flame. The process is used without the application of pressure. This is an obsolete or seldom used process.

air cap. A nonstandard term for the nozzle of a flame spraying gun for wire or ceramic rod.

air carbon arc cutting (CAC-A). A carbon arc cutting process variation that removes molten metal with a jet of air.

air carbon arc cutting torch. A device used to transfer current to a fixed cutting electrode, position the electrode, and direct the flow of air.

air feed. A thermal spraying process variation in which an air stream carries the powdered surfacing material through the gun and into the heat source.

aligned discontinuities. Three or more discontinuities aligned approximately parallel to the weld axis, spaced sufficiently close together to be considered a single intermittent discontinuity.

alloy. A substance with metallic properties and composed of two or more chemical elements of which at least one is a metal.

alloy powder. Powder prepared from a homogeneous molten alloy or from the solidification product of such an alloy. See also **powder blend**.

angle of bevel. See **bevel angle**.

arc. See **welding arc**.

arc blow. The deflection of an arc from its normal path because of magnetic forces.

arc braze welding (ABW). A braze welding process variation that uses an electric arc as the heat source. See also **carbon arc braze welding**.

arc chamber. A nonstandard term for **plenum chamber**.

arc cutter. See **thermal cutter**.

arc cutting (AC). A group of thermal cutting processes that severs or removes metal by melting with the heat of an arc between an electrode and the workpiece.

arc cutting gun. A device used to transfer current to a continuously fed cutting electrode, guide the electrode, and direct the shielding gas.

arc cutting operator. See **thermal cutting operator**.

arc cutting torch. See **air carbon arc cutting torch**, **gas tungsten arc cutting torch**, and **plasma arc cutting torch**.

arc force. The axial force developed by an arc plasma.

arc gap. A nonstandard term when used for **arc length**.

arc gas. A nonstandard term when used for **orifice gas**.

arc gouging. Thermal gouging that uses an arc cutting process variation to form a bevel or groove.

arc length. The distance from the tip of the welding electrode to the adjacent surface of the weld pool.

arc oxygen cutting. A nonstandard term for **oxygen arc cutting**.

arc plasma. A gas that has been heated by an arc to at least a partially ionized condition, enabling it to conduct an electric current.

arc seam weld. A seam weld made by an arc welding process. See Figures 14(A) and (B).

arc seam weld size. See **seam weld size.**

arc spot weld. A spot weld made by an arc welding process. See Figure 14(F).

arc spot weld size. See **spot weld size.**

arc sprayer. See **thermal sprayer.**

arc spraying (ASP). A thermal spraying process using an arc between two consumable electrodes of surfacing materials as a heat source and a compressed gas to atomize and propel the surfacing material to the substrate.

arc spraying operator. See **thermal spraying operator.**

arc strike. A discontinuity resulting from an arc, consisting of any localized remelted metal, heat-affected metal, or change in the surface profile of any metal object.

arc stud welding (SW). An arc welding process that uses an arc between a metal stud, or similar part, and the other workpiece. The process is used without filler metal, with or without shielding gas or flux, with or without partial shielding from a ceramic or graphite ferrule surrounding the stud, and with the application of pressure after the faying surfaces are sufficiently heated.

arc time. The time during which an arc is maintained in making an arc weld.

arc voltage. The voltage across the welding arc.

arc welding (AW). A group of welding processes that produces coalescence of workpieces by heating them with an arc. The processes are used with or without the application of pressure and with or without filler metal.

arc welding deposition efficiency. The ratio of the weight of filler metal deposited in the weld metal to the weight of filler metal melted, expressed in percent.

arc welding electrode. A component of the welding circuit through which current is conducted and that terminates at the arc.

arc welding gun. A device used to transfer current to a continuously fed consumable electrode, guide the electrode, and direct the shielding gas. See Figure 38.

arc welding torch. A device used to transfer current to a fixed welding electrode, position the electrode, and direct the flow of shielding gas. See Figures 35 and 36.

arm, *resistance welding*. A projecting beam extending from the frame of a resistance welding machine that transmits the electrode force and may conduct the welding current.

as-brazed, *adj.* pertaining to the condition of brazements after brazing, prior to any subsequent thermal, mechanical, or chemical treatments.

assist gas. A gas used to blow molten metal away to form the kerf in laser beam inert gas cutting, or to blow vaporized metal away from the beam path in laser beam evaporative cutting.

as-welded, *adj.* pertaining to the condition of weld metal, welded joints, and weldments after welding, but prior to any subsequent thermal, mechanical, or chemical treatments.

atomic hydrogen welding (AHW). An arc welding process that uses an arc between two metal electrodes in a shielding atmosphere of hydrogen and without the application of pressure. This is an obsolete or seldom used process.

autogenous weld. A fusion weld made without filler metal.

| **automatic,** *adj.* pertaining to the control of a process with equipment that requires only occasional or no observation of the welding, and no manual adjustment of the equipment controls.

automatic arc welding current. The current in the welding circuit during the making of a weld, but excluding upslope, downslope, and crater fill current. See Figures 52 and 53.

automatic arc welding downslope time. The time during which the current is changed continuously from final taper current or welding current to final current. See Figure 52.

automatic arc welding upslope time. The time during which the current changes continuously from the initial current to the welding current. See Figure 52.

automatic arc welding weld time. The time interval from the end of start time or end of upslope to beginning of crater fill time or beginning of downslope. See Figures 52 and 53.

automatic brazing. See **automatic welding**.

automatic gas cutting. A nonstandard term for **automatic oxygen cutting**.

automatic soldering. See **automatic welding**.

automatic thermal cutting. See **automatic welding**.

automatic thermal spraying. See **automatic welding**.

automatic welding. Welding with equipment that requires only occasional or no observation of the welding, and no manual adjustment of the equipment controls. Variations of this term are **automatic brazing, automatic soldering, automatic thermal cutting,** and **automatic thermal spraying**. See Figures 52 and 53 and Table 4. See also **adaptive control welding, manual welding, mechanized welding, robotic welding,** and **semiautomatic welding**.

auxiliary enlarger. A nonstandard term for **auxiliary magnifier**.

auxiliary magnifier. An additional lens used to magnify the field of vision.

axis of weld. See **weld axis**.

B

back bead. A weld bead resulting from a back weld pass.

back cap. A device used to exert pressure on the collet in a gas tungsten arc welding torch and create a seal to prevent air from entering the back of the torch. See Figure 36. ||

backfire. The momentary recession of the flame into the welding tip, cutting tip, or flame spraying gun, followed by immediate reappearance or complete extinction of the flame, accompanied by a loud report. |

backgouging. The removal of weld metal and base metal from the weld root side of a welded joint to facilitate complete fusion and complete joint penetration upon subsequent welding from that side.

backhand welding. A welding technique in which the welding torch or gun is directed opposite to the progress of welding. See Figure 21. See also **travel angle, work angle**, and **drag angle**.

backing. A material or device placed against the back side of the joint, or at both sides of a weld in electroslag and electrogas welding, to support and retain molten weld metal. The material may be partially fused or remain unfused during welding and may be either metal or nonmetal. See Figures 8(D) and 12.

backing bead. A weld bead resulting from a backing pass.

backing filler metal. A nonstandard term for **consumable insert**.

backing pass. A weld pass made for a backing weld.

backing ring. Backing in the form of a ring, generally used in the welding of pipe.

backing shoe. A backing device used in electroslag and electrogas welding that remains unfused during welding. See Figure 37.

backing weld. Backing in the form of a weld. See Figure 24(D).

backstep sequence. A longitudinal sequence in which weld passes are made in the direction opposite to the progress of welding. See Figure 23(A).

backup, *flash and upset welding.* A locator used to transmit all or a portion of the upset force to the workpieces or to aid in preventing the workpieces from slipping during upsetting.

back weld. A weld made at the back of a single groove weld. See Figure 24(C).

balling up. The formation of globules of molten filler metal or flux due to lack of wetting of the base metal.

bare electrode. A filler metal electrode that has been produced as a wire, strip, or bar with no coating or covering other than that incidental to its manufacture or preservation.

bare metal arc welding (BMAW). An arc welding process that uses an arc between a bare or lightly coated electrode and the weld pool. The process is used without shielding, without the application of pressure, and filler metal is obtained from the electrode. This is an obsolete or seldom used process.

base material. The material that is welded, brazed, soldered, or cut. See also **base metal** and **substrate.**

base metal. The metal or alloy that is welded, brazed, soldered, or cut. See also **base material** and **substrate.**

base metal test specimen. A test specimen composed wholly of base metal.

base plate. A nonstandard term when used for base metal plate.

bead. See **weld bead.**

bead weld. A nonstandard term for **surfacing weld.**

beam divergence. The expansion of a beam's cross section as the beam emanates from its source.

bend test. A test in which a specimen is bent to a specified bend radius. See also **face bend test, root bend test,** and **side bend test.**

berry formation. A nonstandard term for **nozzle accumulation.**

bevel. An angular edge shape. See Figure 6.

bevel angle. The angle between the bevel of a joint member and a plane perpendicular to the surface of the member. See Figure 6.

bevel-groove weld. A type of groove weld. See Figures 8(B) and 9(B).

bit. That part of the soldering iron, usually made of copper, that directly transfers heat (and sometimes solder) to the joint.

blacksmith welding. A nonstandard term when used for **forge welding.**

blasting. See **abrasive blasting.**

blind joint. A joint, no portion of which is visible.

block brazing (BB). A brazing process that uses heat from heated blocks applied to the joint. This is an obsolete or seldom used process.

block sequence. A combined longitudinal and cross-sectional sequence for a continuous multiple-pass weld in which separated increments are completely or partially welded before intervening increments are welded. See Figure 23(B). See also **progressive block sequence** and **selective block sequence.**

blowhole. A nonstandard term when used for **porosity.**

blowpipe. See **brazing blowpipe** and **soldering blowpipe.**

bond. See **covalent bond, ionic bond, mechanical bond,** and **metallic bond.**

bond bar. A nonstandard term for **bond specimen.**

bond cap. A nonstandard term for **bond specimen.**

bond coat, *thermal spraying.* A preliminary (or prime) coat of material that improves adherence of the subsequent thermal spray deposit.

bonding. A nonstandard term when used for welding, brazing, or soldering.

bonding force. The force that holds two atoms together; it results from a decrease in energy as two atoms are brought closer to one another.

bond line, *thermal spraying.* The cross section of the interface between a thermal spray deposit and the substrate. See Figure 31(B).

bond specimen, *thermal spraying.* The test specimen on which a thermal spray deposit has been applied to determine bond strength and thermal spray deposit strength.

bond strength, *thermal spraying.* The unit force required to separate a thermal spray deposit from the substrate.

bottle. A nonstandard term when used for **gas cylinder**.

boxing. The continuation of a fillet weld around a corner of a member as an extension of the principal weld. See Figure 23(F).

braze. A weld produced by heating an assembly to the brazing temperature using a filler metal having a liquidus above 450°C (840°F) and below the solidus of the base metal. The filler metal is distributed between the closely fitted faying surfaces of the joint by capillary action. See Figure 31(A).

brazeability. The capacity of a material to be brazed under the imposed fabrication conditions into a specific, suitably designed structure, and to perform satisfactorily in the intended service.

braze interface. The interface between braze metal and base metal in a brazed joint. See Figure 31(A).

brazement. An assembly whose component parts are joined by brazing.

braze metal. That portion of a braze that has been melted during brazing.

brazer. One who performs manual brazing.

| **braze welding (BW).** A welding process that uses a filler metal with a liquidus above 450°C (840°F) and below the solidus of the base metal. The base metal is not melted. Unlike brazing, in braze welding the filler metal is not distributed in the joint by capillary action. See also **flow welding.**

| **brazing (B).** A group of welding processes that produces coalescence of materials by heating them to the brazing temperature in the presence of a filler metal having a liquidus above 450°C (840°F) and below the solidus of the base metal. The filler metal is distributed between the closely fitted faying surfaces of the joint by capillary action.

brazing blowpipe. A device used to obtain a small, accurately directed flame for fine work. A portion of any flame is blown to the desired location by the blowpipe, which is usually mouth operated.

| **brazing filler metal.** The metal or alloy used as a filler metal in brazing, which has a liquidus above 450°C (840°F) and below the solidus of the base metal.

brazing operator. One who operates automatic or mechanized brazing equipment.

brazing procedure. The detailed methods and practices involved in the production of a brazement. See also **brazing procedure specification**.

brazing procedure qualification record (BPQR). A record of brazing variables used to produce an acceptable test brazement and the results of tests conducted on the brazement to qualify a brazing procedure specification.

brazing procedure specification (BPS). A document specifying the required brazing variables for a specific application.

brazing sheet. Brazing filler metal in sheet form.

brazing technique. The details of a brazing operation that, within the limitations of the prescribed brazing procedure, are controlled by the brazer or the brazing operator.

brazing temperature. The temperature to which the base metal is heated to enable the filler metal to wet the base metal and form a brazed joint.

brittle nugget. A nonstandard term when used to describe a faying plane failure in a resistance weld peel test.

bronze welding. A nonstandard term when used for **braze welding**.

buildup. A surfacing variation in which surfacing material is deposited to achieve the required dimensions. See also **buttering, cladding**, and **hardfacing**.

buildup sequence. A nonstandard term for **cross-sectional sequence**.

burnback time. A nonstandard term for **meltback time**.

burner. A nonstandard term when used for **oxyfuel gas cutter**.

burning. A nonstandard term when used for **oxyfuel gas cutting**.

burning in. A nonstandard term for **flow welding**.

burnoff rate. A nonstandard term when used for **melting rate**.

burn through. A nonstandard term when used for excessive melt-through or a hole through a root bead.

burn through weld. A nonstandard term for a **seam weld** or **spot weld**.

buttering. A surfacing variation that deposits surfacing metal on one or more surfaces to provide metallurgically compatible weld metal for the subsequent completion of the weld. See also **buildup, cladding,** and **hardfacing**.

butting member. A joint member that is prevented, by the other member, from movement in one direction perpendicular to its thickness dimension. For example, both members of a butt joint, or one member of a T-joint or corner joint. See Figure 11. See also **nonbutting member**.

butt joint. A joint between two members aligned approximately in the same plane. See Figures 1(A), 2(A), 3, 51(A) and 51(B).

button. That part of a weld, including all or part of the nugget, that tears out in the destructive testing of spot, seam, or projection welded specimens.

butt weld. A nonstandard term for a weld in a butt joint.

C

| cap. A nonstandard term for the final layer of a groove weld.

| **capillary action.** The force by which liquid, in contact with a solid, is distributed between closely fitted faying surfaces of the joint to be brazed or soldered.

carbon arc braze welding (CABW). A braze welding process variation that uses an arc between a carbon electrode and the base metal as the heat source.

carbon arc brazing. A nonstandard term for **twin carbon arc brazing.**

carbon arc cutting (CAC). An arc cutting process that uses a carbon electrode. See also **air carbon arc cutting.**

carbon arc welding (CAW). An arc welding process that uses an arc between a carbon electrode and the weld pool. The process is used with or without shielding and without the application of pressure. See also **gas carbon arc welding, shielded carbon arc welding,** and **twin carbon arc welding.**

carbon electrode. A nonfiller metal electrode used in arc welding and cutting, consisting of a carbon or graphite rod, which may be coated with copper or other materials.

carbonizing flame. A nonstandard term for **carburizing flame.**

carburizing flame. A reducing oxyfuel gas flame in which there is an excess of fuel gas, resulting in a carbon-rich zone extending around and beyond the cone. See Figure 40. See also **neutral flame, oxidizing flame,** and **reducing flame.**

carrier gas. The gas used to transport powdered material from the feeder or hopper to a thermal spraying gun or a thermal cutting torch.

cascade sequence. A combined longitudinal and cross-sectional sequence in which weld passes are made in overlapping layers. See Figure 23(C).

caulking. Plastic deformation of weld and adjacent base metal surfaces by mechanical means to seal or obscure discontinuities.

caulk weld. A nonstandard term for **seal weld.**

ceramic rod flame spraying. A thermal spraying process variation in which the surfacing material is in rod form.

chain intermittent weld. An intermittent weld on both sides of a joint in which the weld increments on one side are approximately opposite those on the other side. See Figure 23(G).

chemical flux cutting. A nonstandard term for **flux cutting.**

chill ring. A nonstandard term when used for **backing ring.**

chill time. A nonstandard term when used for **quench time.**

circular electrode. See **resistance welding electrode.**

clad brazing sheet. A metal sheet on which one or both sides are clad with brazing filler metal. See also **clad metal.**

cladding. A surfacing variation that deposits or applies surfacing material usually to improve corrosion or heat resistance. See also **buildup, buttering,** and **hardfacing.**

clad metal. A laminar composite consisting of a metal or alloy, with a metal or alloy of different chemical composition applied to one or more sides by casting, drawing, rolling, surfacing, thick chemical deposition, or thick electroplating.

coalescence. The growing together or growth into one body of the materials being welded.

coated electrode. A nonstandard term for **covered electrode** or **lightly coated electrode.**

coating. A nonstandard term when used for **thermal spray deposit.**

coating density. A nonstandard term when used for **spray deposit density ratio.**

coextrusion welding (CEW). A solid-state welding process that produces a weld by heating to the welding temperature and forcing the workpieces through an extrusion die.

coil without support. A filler metal package consisting of a continuous length of welding wire in coil form without an internal support. It is appropriately bound to maintain its shape.

coil with support. A filler metal package consisting of a continuous length of welding wire in coil form wound on a simple cylinder without flanges. See Figure 42(B).

cold crack. A crack which develops after solidification is complete.

|| cold lap. A nonstandard term when used for **incomplete fusion** or **overlap**.

cold soldered joint. A joint with incomplete coalescence caused by insufficient application of heat to the base metal during soldering.

cold welding (CW). A solid-state welding process in which pressure is used to produce a weld at room temperature with substantial deformation at the weld. See also **diffusion welding**, **forge welding**, and **hot pressure welding**.

collar. The reinforcing metal of a nonpressure thermite weld.

collaring, *thermal spraying*. Adding a shoulder to a shaft or similar component as a protective confining wall for the thermal spray deposit. See Figures 43(A) and (B).

commutator-controlled welding. The making of multiple groups of resistance spot or projection welds sequentially with the same welding contractor through the use of a commutating device.

companion panel. A nonstandard term when used for **spray tab**.

| **complete fusion.** Fusion over the entire fusion faces and between all adjoining weld beads. See Figure 28. See also **incomplete fusion**.

| **complete joint penetration.** A joint root condition in a groove weld in which weld metal extends through the joint thickness. See Figure 26. See also **complete joint penetration weld, incomplete joint penetration, partial joint penetration weld**, and **joint penetration.**

|| **complete joint penetration weld.** A groove weld in which weld metal extends through the joint thickness. See Figure 26. See also **complete joint penetration, incomplete joint penetration, partial joint penetration weld**, and **joint penetration**.

composite. A material consisting of two or more discrete materials with each material retaining its physical identity. See also **clad metal, composite electrode,** and **composite thermal spray deposit**.

composite electrode. A generic term for multicomponent filler metal electrodes in various physical forms such as stranded wires, tubes, and covered wire. See also **covered electrode, flux cored electrode, metal cored electrode,** and **stranded electrode.**

composite thermal spray deposit. A thermal spray deposit made with two or more dissimilar surfacing materials that may be formed in layers.

concave fillet weld. A fillet weld having a concave face. See Figure 25(B).

concave root surface. See Figure 27.

concavity. The maximum distance from the face of a concave fillet weld perpendicular to a line joining the weld toes. See Figure 25(B).

concurrent heating. The application of supplemental heat to a structure during welding or cutting.

cone. The conical part of an oxyfuel gas flame adjacent to the tip orifice. See Figure 40.

connection. A nonstandard term when used for a welded, brazed, or soldered joint.

constant current power source. An arc welding power source with a volt-ampere relationship yielding a small welding current change from a large arc voltage change. See also **welding power source.** ||

constant voltage power source. An arc welding power source with a volt-ampere relationship yielding a large welding current change from a small arc voltage change. See also **welding power source.** ||

constricted arc. A plasma arc column that is shaped by the constricting orifice in the nozzle of the plasma arc torch or plasma spraying gun.

constricting nozzle. A device at the exit end of a plasma arc torch or plasma spraying gun, containing the constricting orifice. See Figure 35. |

constricting orifice. The hole in the constricting nozzle of the plasma arc torch or plasma spraying gun through which the arc plasma passes. See Figure 35. |

constricting orifice diameter. See Figure 35. ||

constricting orifice length. See Figure 35. ||

consumable electrode. An electrode that provides filler metal.

consumable guide electroslag welding. An electroslag welding process variation in which filler metal is supplied by an electrode and its guiding member.

consumable insert. Filler metal that is placed at the joint root before welding, and is intended to be completely fused into the joint root to become part of the weld. |

contact resistance, *resistance welding.* Resistance to the flow of electric current between two workpieces or an electrode and a workpiece.

contact tube. A device that transfers current to a continuous electrode. See Figures 38 and 39.

contact tube setback, *flux cored arc welding and gas metal arc welding.* The distance from the contact tube to the end of the gas nozzle. See Figure 38. See also **electrode setback.**

continuous sequence. A longitudinal sequence in which each weld pass is made continuously from one end of the weld to the other.

continuous wave laser. A laser having an output that operates in a continuous rather than a pulsed mode. A laser operating with a continuous output for a period greater than 25 milliseconds is regarded as a continuous wave laser.

continuous weld. A weld that extends continuously from one end of a joint to the other. Where the joint is essentially circular, it extends completely around the joint.

convex fillet weld. A fillet weld having a convex face. See Figure 25(A).

convexity. The maximum distance from the face of a convex fillet weld perpendicular to a line joining the weld toes. See Figure 25(A).

convex root surface. See Figure 27.

cool time, *resistance welding.* The time interval between successive heat times in multiple-impulse welding or in the making of seam welds. See Figures 48(B) and 49.

copper brazing. A nonstandard term when used for brazing with a copper filler metal.

cord, *thermal spraying.* Surfacing material in the form of a plastic tube filled with powder that has been extruded to a compact, flexible cord with characteristics similar to a wire.

cored solder. A solder wire or bar containing flux as a core.

corner-flange weld. A nonstandard term for an edge weld in a flanged corner joint.

corner joint. A joint between two members located approximately at right angles to each other in the form of an L.

corona, *resistance welding.* The area sometimes surrounding the nugget of a spot weld at the faying surfaces which provides a degree of solid-state welding.

corrective lens. A lens ground to the wearer's individual corrective prescription.

corrosive flux. A flux with a residue that chemically attacks the base metal. It may be composed of inorganic salts and acids, organic salts and acids, or activated rosin.

cosmetic pass. A weld pass made primarily to enhance appearance.

CO_2 welding. A nonstandard term for gas metal arc welding with carbon dioxide shielding gas.

covalent bond. A primary bond arising from the reduction in energy associated with overlapping half-filled orbitals of two atoms.

covered electrode. A composite filler metal electrode consisting of a core of a bare electrode or metal cored electrode to which a covering sufficient to provide a slag layer on the weld metal has been applied. The covering may contain materials providing such functions as shielding from the atmosphere, deoxidation, and arc stabilization, and can serve as a source of metallic additions to the weld.

cover lens. A nonstandard term for a round cover plate.

cover plate. A removable pane of colorless glass, plastic-coated glass, or plastic that covers the filter plate and protects it from weld spatter, pitting, or scratching.

crack. A fracture type discontinuity characterized by a sharp tip and high ratio of length and width to opening displacement. See Figure 33.

crater. A depression in the weld face at the termination of a weld bead.

crater crack. See Figure 33.

crater fill current. The current value during crater fill time. See Figure 53.

crater fill time. The time interval following weld time but prior to meltback time during which arc voltage or current reach a preset value greater or less than welding values. Weld travel may or may not stop at this point. See Figure 53.

crater fill voltage. The arc voltage value during crater fill time. See Figure 53.

cross-sectional sequence. The order in which the weld passes of a multiple-pass weld are made with respect to the cross section of the weld. See Figures 23(D) and (E). See also **block sequence** and **cascade sequence.**

cross wire welding. A common variation of projection welding wherein the localization of the welding current is achieved by the intersection contact of wires, and is usually accompanied by considerable embedding of one wire into another.

cup. A nonstandard term when used for **gas nozzle**.

cutter. See **thermal cutter**.

cutting. See **thermal cutting**.

cutting attachment. A device for converting an oxyfuel gas welding torch into an oxygen cutting torch.

cutting blowpipe. A nonstandard term for **oxyfuel gas cutting torch**.

cutting electrode. A nonfiller metal electrode used in arc cutting. See also **carbon electrode, metal electrode**, and **tungsten electrode**.

cutting head. The part of a cutting machine in which a cutting torch or tip is incorporated.

cutting nozzle. A nonstandard term for **cutting tip**.

cutting operator. See **thermal cutting operator**.

cutting tip. The part of an oxygen cutting torch from which the gases issue. See Figure 41.

| **cutting torch.** See **air carbon arc cutting torch, gas tungsten arc cutting torch**, and **plasma arc cutting torch**, and **oxyfuel gas cutting torch**.

|| **cycle.** The duration of alternating current represented by the current increase from an initial value to a maximum in one direction then to a maximum in the reverse direction and its return to the original initial value.

cylinder. See **gas cylinder**.

cylinder manifold. A multiple header for interconnection of gas sources with distribution points.

D

defect. A discontinuity or discontinuities that by nature or accumulated effect (for example total crack length) render a part or product unable to meet minimum applicable acceptance standards or specifications. The term designates rejectability. See also **discontinuity** and **flaw**.

delayed cracking. A nonstandard term when used for cold cracking caused by hydrogen embrittlement.

deposit. A nonstandard term when used for **thermal spray deposit**.

|| **deposited metal,** *welding, brazing, and soldering*. Filler metal that has been added during welding, brazing or soldering.

deposited metal, *surfacing*. Surfacing metal that has been added during surfacing. ||

deposition efficiency. See arc **welding deposition efficiency** and **thermal spraying deposition efficiency**.

deposition rate. The weight of material deposited in a unit of time.

deposition sequence. A nonstandard term when used for **weld pass sequence**.

deposit sequence. A nonstandard term when used for **weld pass sequence**.

depth of bevel. The perpendicular distance from the base metal surface to the root edge or the beginning of the root face. See Figure 6.

depth of fusion. The distance that fusion extends into the base metal or previous bead from the surface melted during welding. See Figure 30. See also **joint penetration**.

detonation flame spraying. A thermal spraying process variation in which the controlled explosion of a mixture of fuel gas, oxygen, and powdered surfacing material is utilized to melt and propel the surfacing material to the substrate.

die. See **resistance welding die** and **forge welding die**.

die welding. A nonstandard term when used for **cold welding** and **forge welding**.

diffusion aid. A solid filler metal applied to the faying surfaces to assist in diffusion welding.

diffusion bonding. A nonstandard term for **diffusion brazing** and **diffusion welding**.

diffusion brazing (DFB). A brazing process that produces coalescence of metals by heating them to brazing temperature and by using a filler metal or an *in situ* liquid phase. The filler metal may be distributed by capillary attraction or may be placed or formed at the faying surfaces. The filler metal is diffused with the base metal to the extent that the joint properties have been changed to approach those of the base metal. Pressure may or may not be applied.

diffusion welding (DFW). A solid-state welding process that produces a weld by the application of pressure at elevated temperature with no macroscopic deformation or relative motion of the workpieces. A solid filler metal may be inserted between the faying surfaces. See also **cold welding, diffusion aid, forge welding**, and **hot pressure welding**.

dilution. The change in chemical composition of a welding filler metal caused by the admixture of the base metal or previous weld metal in the weld bead. It is measured by the percentage of base metal or previous weld metal in the weld bead. See Figure 24(L).

dip brazing (DB). A brazing process that uses heat from a molten salt or metal bath. When a molten salt is used, the bath may act as a flux. When a molten metal is used, the bath provides the filler metal. See also **metal-bath dip brazing** and **salt-bath dip brazing.**

dip soldering (DS). A soldering process using the heat furnished by a molten metal bath that provides the solder filler metal.

dip transfer. A nonstandard term for **short circuiting transfer.**

direct current electrode negative (DCEN). The arrangement of direct current arc welding leads in which the electrode is the negative pole and workpiece is the positive pole of the welding arc. See Figure 34(B).

direct current electrode positive (DCEP). The arrangement of direct current arc welding leads in which the electrode is the positive pole and the workpiece is the negative pole of the welding arc. See Figure 34(A).

direct current reverse polarity. A nonstandard term for **direct current electrode positive.**

direct current straight polarity. A nonstandard term for **direct current electrode negative.**

direct drive friction welding. A variation of friction welding in which the energy required to make the weld is supplied to the welding machine through a direct motor connection for a preset period of the welding cycle. See Figure 45.

direct welding, *resistance welding.* A resistance welding secondary circuit variation in which welding current and electrode force are applied to the workpieces by directly opposed electrodes, wheels, or conductor bars for spot, seam, or projection welding. See Figures 47(A), 47(B), and 47(C).

discontinuity. An interruption of the typical structure of a material, such as a lack of homogeneity in its mechanical, metallurgical, or physical characteristics. A discontinuity is not necessarily a defect. See also **defect** and **flaw.**

| **doped solder.** A solder containing a small amount of an element added to ensure retention of one or more characteristics of the base materials on which it is used.

double arcing. A condition in which the welding or cutting arc of a plasma arc torch does not pass through the constricting orifice but transfers to the inside surface of the nozzle. A secondary arc is simultaneously established between the outside surface of the nozzle and the workpiece.

double-bevel edge shape. A type of edge shape. See Figure 7(C).

double-bevel-groove weld. A type of groove weld. See Figure 9(B).

double-flare-bevel-groove weld. A weld in grooves formed by a member with a curved surface in contact with a planar member. See Figure 9(F).

double-flare-V-groove weld. A weld in grooves formed by two members with curved surfaces. See Figure 9(G).

double-groove weld, *fusion welding.* A groove weld that is made from both sides. See Figures 9, 24(C), and 24(D). ||

double-J edge shape. A type of edge shape. See Figure 7(E).

double-J-groove weld. A type of groove weld. See Figure 9(D).

double-square-groove weld. A type of groove weld. See Figure 9(A).

double-U-groove weld. A type of groove weld. See Figure 9(E).

double-V-groove weld. A type of groove weld. See Figure 9(C).

double-spliced butt joint. See **spliced joint.** See Figure 3. ||

double-welded joint, *fusion welding.* A joint that is welded from both sides. See Figures 9, 24(C), and 24(D). ||

dovetailing, *thermal spraying.* A method of surface roughening involving angular undercutting to interlock the thermal spray deposit. See Figure 43(C).

downhand. A nonstandard term for **flat welding position.** |

downhill, *adv.* Welding with a downward progression.

downslope time. See **automatic arc welding downslope time** and **resistance welding downslope time.**

drag, *thermal cutting.* The offset distance between the actual and straight line exit points of the gas stream or cutting beam measured on the exit surface of the base metal. See Figure 41.

drag angle. The travel angle when the electrode is pointing in a direction opposite to the progression of welding. This angle can also be used to partially define the position of guns, torches, rods, and beams. See Figure 21. See also **backhand welding, push angle, travel angle,** and **work angle.**

drop-through. An undesirable sagging or surface irregularity, usually encountered when brazing or welding near the solidus of the base metal, caused by overheating with rapid diffusion or alloying between the filler metal and the base metal.

drum. A filler metal package consisting of a continuous length of welding wire wound or coiled in a cylindrical container.

duty cycle. The percentage of time during an arbitrary test period that a power source or its accessories can be operated at rated output without overheating.

dwell time, *thermal spraying.* The length of time that the surfacing material is exposed to the heat zone of the thermal spraying gun.

dynamic electrode force. The force exerted by electrodes on the workpieces during the actual welding cycle in making spot, seam, or projection welds by resistance welding.

E

edge effect, *thermal spraying.* Loosening of the bond between the thermal spray deposit and the substrate at the edge of the thermal spray deposit.

edge-flange weld. A nonstandard term for an edge weld in a flanged butt joint.

edge joint. A joint between the edges of two or more parallel or nearly parallel members. See Figures 1(E) and 2(E).

edge loss, *thermal spraying.* Thermal spray deposit lost as overspray beyond the edge of the workpiece.

edge preparation. The preparation of the edges of the joint members, by cutting, cleaning, plating, or other means.

edge preparation. A nonstandard term when used for edge shape.

edge shape. The shape of the edge of the joint member. See Figure 7.

edge weld. A weld in an edge joint, a flanged butt joint or a flanged corner joint in which the full thickness of the members are fused. See Figures 10(A), 10(B), 10(C), 13(A), and 25(H).

edge weld size. The weld metal thickness measured from the weld root. See Figure 25(H).

effective throat. The minimum distance minus any convexity between the weld root and the face of a fillet weld. See Figure 25(A), (B), (C), and (D).

electric arc spraying. A nonstandard term for **arc spraying.**

electric bonding. A nonstandard term for surfacing by thermal spraying.

electric brazing. A nonstandard term for **arc brazing** and **resistance brazing.**

electrode. A component of the electrical circuit that terminates at the arc, molten conductive slag, or base metal. See **cutting electrode, tungsten electrode,** and **welding electrode.**

electrode cap. A replaceable electrode tip used for resistance spot welding.

electrode extension, *carbon arc cutting.* The length of electrode extending beyond the electrode holder or cutting torch.

electrode extension, *flux cored arc welding, electrogas welding, gas metal arc welding, and submerged arc welding.* The length of electrode extending beyond the end of the contact tube. See Figure 38.

electrode extension, *gas tungsten arc welding and plasma arc welding.* The length of tungsten electrode extending beyond the end of the collet. See Figure 36.

electrode force. The force applied to the electrodes in making spot, seam, or projection welds by resistance welding. See also **dynamic electrode force, static electrode force,** and **theoretical electrode force.**

electrode gap. A nonstandard term for **arc length.**

electrode holder. A device used for mechanically holding and conducting current to an electrode during welding or cutting. See Figure 34.

electrode indentation, *resistance welding.* The depression formed on the surface of workpieces by electrodes.

electrode lead. The electrical conductor between the source of arc welding current and the electrode holder. See Figures 34 and 36.

electrode mushrooming. The enlargement of a resistance spot or projection welding electrode tip due to heat or pressure so it resembles a mushroom in shape.

electrode pickup. Contamination of the electrode tips or wheel faces by the base metal or its coating during resistance spot, seam, or projection welding.

electrode setback. The distance the electrode is recessed behind the constricting orifice of the plasma arc torch or thermal spraying gun, measured from the outer face of the constricting nozzle. See Figure 35. See also **contact tube setback.**

electrode skid. The sliding of a resistance welding electrode along the surface of the workpiece when making spot, seam, or projection welds.

electrode tip. The end of a resistance spot or projection welding electrode in contact with the workpiece.

electrode tip life. The number of resistance spot welds that can be made with an electrode before redressing of the electrode is required.

electrogas welding (EGW). An arc welding process that uses an arc between a continuous filler metal electrode and the weld pool, employing approximately vertical welding progression with backing to confine the molten weld metal. The process is used with or without an externally supplied shielding gas and without the application of pressure.

electron beam braze welding (EBBW). A braze welding process variation that uses an electron beam as the heat source.

electron beam cutting (EBC). A thermal cutting process that severs metals by melting them with the heat from a concentrated beam, composed primarily of high-velocity electrons, impinging on the workpiece.

electron beam cutting operator. See **thermal cutting operator**.

electron beam gun. A device for producing and accelerating electrons. Typical components include the emitter (also called the filament or cathode) that is heated to produce electrons via thermionic emission, a cup (also called the grid or grid cup), and the anode.

electron beam gun column. The electron beam gun plus auxiliary mechanical and electrical components that may include beam alignment, focus, and deflection coils.

electron beam welding (EBW). A welding process that produces coalescence with a concentrated beam, composed primarily of high-velocity electrons, impinging on the joint. The process is used without shielding gas and without the application of pressure. See also **high vacuum electron beam welding, medium vacuum electron beam welding**, and **nonvacuum electron beam welding**.

electroslag welding (ESW). A welding process that produces coalescence of metals with molten slag that melts the filler metal and the surfaces of the workpieces. The weld pool is shielded by this slag, which moves along the full cross section of the joint as welding progresses. The process is initiated by an arc that heats the slag. The arc is then extinguished by the conductive slag, which is kept molten by its resistance to electric current passing between the electrode and the workpieces. See also **electroslag welding electrode** and **consumable guide electroslag welding**.

electroslag welding electrode. A filler metal component of the welding circuit through which current is conducted from the electrode guiding member to the molten slag.

emissive electrode. A filler metal electrode consisting of a core of a bare electrode or a composite electrode to which a very light coating has been applied to produce a stable arc.

end return. A nonstandard term for **boxing**.

erosion, *brazing*. A condition caused by dissolution of the base metal by molten filler metal resulting in a reduction in the thickness of the base metal.

exhaust booth. A mechanically ventilated, semi-enclosed area in which an air flow across the work area is used to remove fumes, gases, and solid particles.

exothermic braze welding (EXBW). A braze welding process variation that uses an exothermic chemical reaction between a metal oxide and a metal or inorganic nonmetal as the heat source, with a reaction product as the filler metal.

exothermic brazing (EXB). A brazing process using an exothermic chemical reaction between a metal oxide and a metal or inorganic nonmetal as the heat source, with filler metal preplaced in the joint.

explosion welding (EXW). A solid-state welding process that produces a weld by high velocity impact of the workpieces as the result of controlled detonation.

expulsion. The forceful ejection of molten metal from a resistance spot, seam, or projection weld usually at the faying surface. See also **surface expulsion**.

expulsion point. The level of welding current (for a specific set of welding conditions in spot, seam, or projection welding) above which there is significant expulsion.

extension. The distance a workpiece or electrode projects from a welding die, clamp, chuck, or holder.

F

face bend test. A test in which the weld face is on the convex surface of a specified bend radius.

face crack. See Figure 33.

face feed. The application of filler metal to the joint, usually by hand, during brazing and soldering.

face of weld. See **weld face**.

face reinforcement. Weld reinforcement on the side of the joint from which welding was done. See Figures 24(A) and (C). See also **root reinforcement**.

face shield. A device positioned in front of the eyes and over all or a portion of the face to protect the eyes and face. See also **hand shield** and **helmet**.

faying surface. The mating surface of a member that is in contact with or in close proximity to another member to which it is to be joined. See Figure 30(D).

feather. See **acetylene feather**.

feed rate, *thermal spraying*. A nonstandard term for **spray rate.**

Ferrite Number (FN). An arbitrary, standardized value designating the ferrite content of an austenitic stainless steel weld metal. It should be used in place of percent ferrite or volume percent ferrite on a direct replacement basis.

ferrule, *arc stud welding*. A ceramic device that surrounds the stud base to contain the molten metal and shield the arc.

field weld. A weld made at a location other than a shop or the place of initial construction.

filler. See **joint filler.**

filler material. The material to be added in making a welded, brazed , or soldered joint. See also **brazing filler metal, consumable insert, diffusion aid, filler metal, solder, welding electrode, welding filler metal, welding rod**, and **welding wire**.

filler metal. The metal or alloy to be added in making a welded, brazed, or soldered joint. See also **brazing filler metal, consumable insert, diffusion aid, filler material, solder, welding electrode, welding filler metal, welding rod**, and **welding wire**.

filler metal start delay time. The time interval from arc initiation to the start of filler metal feeding. See Figure 52.

filler metal stop delay time. The time delay interval from beginning of downslope time to the stop of filler metal feeding. See Figure 52.

filler wire. A nonstandard term for **welding wire**.

fillet weld. A weld of approximately triangular cross section joining two surfaces approximately at right angles to each other in a lap joint, T-joint, or corner joint. See Figures 10(F), 15(F), 18, 20, 23(G), 23(H), 24(E), 24(J), and 25(A)–(E).

fillet weld break test. A test in which the specimen is loaded so that the weld root is in tension.

fillet weld leg. The distance from the joint root to the toe of the fillet weld. See Figures 24(E) and 25(A)–(E).

fillet weld size. For equal leg fillet welds, the leg lengths of the largest isosceles right triangle that can be inscribed within the fillet weld cross section. For unequal leg fillet welds, the leg lengths of the largest right triangle that can be inscribed within the fillet weld cross section. See Figures 25(A)–(E).

fillet weld throat. See **actual throat, effective throat,** and **theoretical throat.**

fill weld. A fusion weld made with filler metal.

filter glass. A nonstandard term for **filter plate.**

filter lens. A nonstandard term for a round filter plate.

filter plate. An optical material that protects the eyes against excessive ultraviolet, infrared, and visible radiation.

final current. The current after downslope but prior to current shut-off. See Figure 52.

final taper current. The current at the end of the taper interval prior to downslope. See Figure 52.

fines. All material finer than a particular mesh size under consideration.

firecracker welding. A shielded metal arc welding process variation that uses a length of covered electrode placed along the joint in contact with the workpieces during welding. The stationary electrode is consumed as the arc travels the length of the electrode. This is an obsolete or seldom used process variation.

fisheye. A discontinuity found on the fracture surface of a weld in steel that consists of a small pore or inclusion surrounded by an approximately round, bright area.

5F. A welding test position designation for a circumferential fillet weld applied to a joint in pipe, with its axis approximately horizontal, in which the weld is made in the horizontal, vertical, and overhead welding positions. The pipe remains fixed until the welding of the joint is complete. See Figure 20(E).

5G. A welding test position designation for a circumferential groove weld applied to a joint in a pipe with its axis horizontal, in which the weld is made in the flat, vertical, and overhead welding positions. The pipe remains fixed until the welding of the joint is complete. See Figure 19(C).

fixture. A device designed to hold and maintain parts in proper relation to each other.

flame. See **carburizing flame, neutral flame, oxidizing flame**, and **reducing flame**.

flame cutting. A nonstandard term for **oxygen cutting.**

flame propagation rate. The speed at which flame travels through a mixture of gases.

flame sprayer. See **thermal sprayer.**

flame spraying (FLSP). A thermal spraying process in which an oxyfuel gas flame is the source of heat for melting the surfacing material. Compressed gas may or may not be used for atomizing and propelling the surfacing material to the substrate.

flame spraying operator. See **thermal spraying operator.**

|| **flanged butt joint.** A form of a butt joint in which at least one of the members has a flanged edge shape at the joint. See Figures 2(A), 10(A), 10(B), 10(D) and 27(D).

|| **flanged corner joint.** A form of a corner joint in which the butting member has a flanged edge shape at the joint, and an edge weld is applicable. See Figures 2(B), 10(C), 10(E), and 27(B).

flanged edge joint. A form of an edge joint in which at least one of the members has a flanged edge shape at the joint. See Figure 2(E).

flanged edge shape. A type of edge shape produced by forming the member. See Figure 7(F).

|| **flanged joint.** A form of one of the five basic joint types in which at least one of the joint members has a flanged edge shape at the weld joint. See Figures 2, 10 and 27(B) and (D).

flanged lap joint. A form of a lap joint in which at least one of the members has a flanged edge shape at the joint, and an edge weld is not applicable. See Figure 2(D).

flanged T-joint. A form of a T-joint in which the butting member has a flanged edge shape at the joint, and an edge weld is not applicable. See Figures 2(C) and 10(F).

flange weld. A nonstandard term for a weld in a flanged joint.

| **flare-bevel-groove weld.** A weld in the groove formed between a joint member with a curved surface and another with a planar surface. See Figures 8(G), 9(F), 10(F) and 26(H).

|| **flare-groove weld.** A weld in the groove formed between a joint member with a curved surface and another with a planar surface, or between two joint members with curved surfaces. See Figures 8(G), 8(H), 9(F), 9(G), 10(D), and 10(F). See also **flare-bevel-groove weld** and **flare-V-groove weld.**

| **flare-V-groove weld.** A weld in a groove formed by two members with curved surfaces. See Figures 8(H), 9(G) and 10(D).

flash. Material that is expelled from a flash weld prior to the upset portion of the welding cycle.

flashback. A recession of the flame into or back of the mixing chamber of the oxyfuel gas torch or flame spraying gun.

flashback arrester. A device to limit damage from a flashback by preventing propagation of the flame front beyond the location of the arrester.

flash butt welding. A nonstandard term for **flash welding.**

flash coat. A thin coating usually less than 0.002 in. in thickness.

flashing action. The phenomenon in flash welding where in points of contact, formed by light pressure across faying surfaces, are melted and explosively ejected because of the extremely high current density at contact points.

flash off time. A nonstandard term for **flash time.**

flash time. The duration of flashing action during flash welding.

flash welding (FW). A resistance welding process that produces a weld at the faying surfaces of a butt joint by a flashing action and by the application of pressure after heating is substantially completed. The flashing action, caused by the very high current densities at small contact points between the workpieces, forcibly expels the material from the joint as the workpieces are slowly moved together. The weld is completed by a rapid upsetting of the workpieces. See Figure 15(B).

flat position. See **flat welding position.** |

flat welding position. The welding position used to weld from the upper side of the joint at a point where the weld axis is approximately horizontal, and the weld face lies in an approximately horizontal plane. See Figures 16A, 16B, 16C, 17(A), 18(A), 19(A), and 20(A). ||

flaw. An undesirable discontinuity. See also **defect.**

flood cooling, *resistance seam welding.* The application of liquid coolant directly on the work and the contacting electrodes.

flowability. The ability of molten filler metal to flow or spread over a metal surface.

flow brazing (FLB). A brazing process that uses heat from molten nonferrous filler metal poured over the joint until brazing temperature is attained. This is an obsolete or seldom used process.

flow brightening, *soldering.* Fusion of a metallic coating on a base metal.

flow welding (FLOW). A braze welding process variation that uses molten filler metal poured over the fusion faces as the heat source. This is an obsolete or seldom used process.

flux. A material used to hinder or prevent the formation of oxides and other undesirable substances in molten metal and on solid metal surfaces, and to dissolve or otherwise facilitate the removal of such substances. See also **active flux** and **neutral flux**.

flux cored arc welding (FCAW). An arc welding process that uses an arc between a continuous filler metal electrode and the weld pool. The process is used with shielding gas from a flux contained within the tubular electrode, with or without additional shielding from an externally supplied gas, and without the application of pressure. See also **flux cored electrode**, **gas shielded flux cored arc welding**, and **self-shielded flux cored arc welding**.

flux cored electrode. A composite tubular filler metal electrode consisting of a metal sheath and a core of various powdered materials, producing an extensive slag cover on the face of a weld bead. External shielding may be required.

flux cover, *metal bath dip brazing and dip soldering.* A layer of molten flux over the molten filler metal bath.

flux cutting (FOC). An oxygen cutting process that uses heat from an oxyfuel gas flame, with a flux in the flame to aid cutting.

flux oxygen cutting. A nonstandard term for **flux cutting**.

focal point. A nonstandard term for **focal spot**.

focal spot, *electron beam welding and cutting, and laser beam welding and cutting.* A location at which the beam has the most concentrated energy and the smallest cross-sectional area.

follow-up, *resistance welding.* The ability of the moveable electrode to maintain proper electrode force and contact with the workpiece as metal movement occurs, especially in projection welding.

forehand welding. A welding technique in which the welding torch or gun is directed toward the progress of welding. See Figure 21. See also **travel angle**, **work angle**, and **push angle**.

forge-delay time, *resistance welding.* The time elapsing between a preselected point in the welding cycle and the initiation of the forging force. See Figure 49.

forge force. A compressive force applied to the weld after the heating portion of the welding cycle is essentially complete.

forge welding (FOW). A solid-state welding process that produces a weld by heating the workpieces to welding temperature and applying blows sufficient to cause permanent deformation at the faying surfaces. See also **cold welding**, **diffusion welding**, and **hot pressure welding**.

forging speed, *friction welding.* The relative velocity of the workpieces at the instant the forge force is applied.

4F, *plate.* A welding test position designation for a linear fillet weld applied to a joint in which the weld is made in the overhead welding position. See Figure 18(D).

4F, *pipe.* A welding test position designation for a circumferential fillet weld applied to a joint in pipe, with its axis vertical, in which the weld is made in the overhead welding position. See Figure 20(D).

4G. A welding test position designation for a linear groove weld applied to a joint in which the weld is made in the overhead welding position. See Figure 17(D).

friction soldering. A nonstandard term for **abrasion soldering**.

friction speed, *friction welding.* The relative velocity of the workpieces at the time of initial contact. See Figures 44 and 45.

friction upset distance. The decrease in length of work pieces during the time of friction welding force application. See Figures 44 and 45.

friction welding (FRW). A solid-state welding process that produces a weld under compressive force contact of workpieces rotating or moving relative to one another to produce heat and plastically displace material from the faying surfaces. See Figures 31(D), 44, and 45. See also **direct drive friction welding** and **inertia friction welding**.

friction welding force. The compressive force applied to the faying surfaces during the time there is relative movement between the workpieces from the start of welding until the application of the forge force. See Figures 44 and 45.

fuel gas. A gas such as acetylene, natural gas, hydrogen, propane, stabilized methylacetylene propadiene, and other fuels normally used with oxygen in one of the oxyfuel processes and for heating.

full fillet weld. A fillet weld equal in size to the thickness of the thinner member joined.

full penetration. A nonstandard term for **complete joint penetration**.

furnace brazing (FB). A brazing process in which the workpieces are placed in a furnace and heated to the brazing temperature.

furnace soldering (FS). A soldering process in which the workpieces are placed in a furnace and heated to the soldering temperature.

fused thermal spray deposit. A self-fluxing thermal spray deposit that is subsequently heated to coalescence within itself and with the substrate.

fused zone. A nonstandard term for **fusion zone.**

fusing. A nonstandard term for **fusion.**

fusion, *fusion welding.* The melting together of filler metal and base metal, or of base metal only, to produce a weld. See also **depth of fusion.**

fusion face. A surface of the base metal that will be melted during welding. See Figure 30.

fusion line. A nonstandard term for **weld interface.**

fusion welding. Any welding process that uses fusion of the base metal to make the weld.

fusion zone. The area of base metal melted as determined on the cross section of a weld. See Figure 30.

G

gap. A nonstandard term when used for **arc length, joint clearance,** and **root opening.**

gas brazing. A nonstandard term for **torch brazing.**

gas carbon arc welding (CAW-G). A carbon arc welding process variation that uses a shielding gas. This is an obsolete or seldom used process.

gas cup. A nonstandard term for **gas nozzle.**

gas cutter. A nonstandard term for **oxygen cutter.**

gas cutting. A nonstandard term for **oxygen cutting.**

gas cylinder. A portable container used for transportation and storage of compressed gas.

gas gouging. A nonstandard term for **oxygen gouging.**

gas laser. A laser in which the lasing medium is a gas.

gas lens. One or more fine mesh screens located in the torch nozzle to produce a stable stream of shielding gas. Primarily used for gas tungsten arc welding.

gas metal arc cutting (GMAC). An arc cutting process that uses a continuous consumable electrode and a shielding gas.

gas metal arc welding (GMAW). An arc welding process that uses an arc between a continuous filler metal electrode and the weld pool. The process is used with shielding from an externally supplied gas and without thc application of pressure. See also **pulsed gas metal arc welding** and **short circuit gas metal arc welding.**

gas nozzle. A device at the exit end of the torch or gun that directs shielding gas. See Figures 35, 36, 38 and 39.

gas pocket. A nonstandard term for **porosity.**

gas regulator. A device for controlling the delivery of gas at some substantially constant pressure.

gas shielded arc welding. A group of processes including **electrogas welding, flux cored arc welding, gas metal arc welding, gas tungsten arc welding,** and **plasma arc welding.**

gas shielded flux cored arc welding (FCAW-G). A flux cored arc welding process variation in which shielding gas is supplied through the gas nozzle, in addition to that obtained from the flux within the electrode.

gas torch. A nonstandard term for **welding torch** and **cutting torch.**

gas tungsten arc cutting (GTAC). An arc cutting process that uses a single tungsten electrode with gas shielding.

gas tungsten arc cutting torch. A device used to transfer current to a fixed cutting electrode, position the electrode, and direct the flow of shielding gas.

gas tungsten arc welding (GTAW). An arc welding process that uses an arc between a tungsten electrode (nonconsumable) and the weld pool. The process is used with shielding gas and without the application of pressure. See also **hot wire welding** and **pulsed gas tungsten arc welding.**

gas tungsten arc welding torch. A device used to transfer current to a fixed welding electrode, position the electrode, and direct the flow of shielding gas. See Figure 36.

gas welding. A nonstandard term for **oxyfuel gas welding.**

globular arc. A nonstandard term for **globular transfer.**

globular transfer, *arc welding.* The transfer of molten metal in large drops from a consumable electrode across the arc. See Figure 39(A). See also **short circuiting transfer** and **spray transfer.**

goggles. Protective glasses equipped with filter plates set in a frame that fits snugly against the face and used primarily with oxyfuel gas processes.

gouging. See **thermal gouging.**

governing metal thickness, *resistance welding.* The thickness of the sheet on which the required weld nugget size and depth of fusion is based.

gradated thermal spray deposit. A composite thermal spray deposit composed of mixed materials in successive layers that progressively change in composition from the material adjacent to the substrate to the material at the surface of the thermal spray deposit.

gravity feed welding. A shielded metal arc welding process variation for making a fillet weld in which a long electrode slides down a tripod-mounted electrode holder as the electrode is consumed.

groove. See **scarf groove** and **weld groove.**

groove and rotary roughening, *thermal spraying.* A method of surface roughening in which grooves are made and the original surface is roughened and spread.

groove angle. The total included angle of the groove between workpieces. See Figure 6.

groove face. The surface of a joint member included in the groove. See Figure 5.

groove radius. The radius used to form the shape of a J- or U-groove weld. See Figure 6.

groove weld. A weld made in a groove between the workpieces. See Figures 8, 9, and 21(A).

groove weld size. The joint penetration of a groove weld. See Figure 26.

groove weld throat. A nonstandard term for **groove weld size.**

ground clamp. A nonstandard and incorrect term for **workpiece connection.**

ground connection. An electrical connection of the welding machine frame to the earth for safety. See Figure 34. See also **workpiece connection** and **workpiece lead.**

ground lead. A nonstandard and incorrect term for **workpiece lead.**

gun. See **arc cutting gun, arc welding gun, electron beam gun, resistance welding gun, soldering gun,** and **thermal spraying gun.**

gun extension. The extension tube attached in front of the thermal spraying gun to permit spraying within confined areas or deep recesses.

H

hammering, *resistance spot welding.* Excessive electrode impact on the surface of the workpiece at the start of the welding cycle.

hammer welding. A nonstandard term for **forge welding** and **cold welding.**

hand shield. A protective device used in arc welding, arc cutting and thermal spraying, for shielding the eyes, face, and neck. It is equipped with a filter plate and is designed to be held by hand.

hardfacing. A surfacing variation in which surfacing material is deposited to reduce wear. See also **buildup, buttering,** and **cladding.**

hard solder. A nonstandard term for brazing filler metal.

hard surfacing. A nonstandard term for **hardfacing.**

head. See **welding head** and **cutting head.**

heat-affected zone. The portion of the base metal whose mechanical properties or microstructure have been altered by the heat of welding, brazing, soldering, or thermal cutting. See Figure 24(G).

heat-affected-zone crack. See Figure 33.

heat balance. The various material, joint, and welding conditions that determine the welding heat pattern in the joint.

heating gate. The opening in a thermite mold through which the workpieces are preheated.

heating torch. A device for directing the heating flame produced by the controlled combustion of fuel gases.

heat time. The duration of any one impulse in multiple impulse welding or resistance seam welding. See Figures 48(B) and 49.

helmet. See **welding helmet.**

high-frequency resistance welding. A group of resistance welding process variations that use high-frequency welding current to concentrate the welding heat at the desired location. See Figure 51. See also **high-frequency seam welding** and **high-frequency upset welding.**

high-frequency seam welding (RSEW-HF). A resistance seam welding process variation in which high-frequency welding current is supplied through electrodes into the workpieces. See Figure 51(C). See also **high-frequency resistance welding** and **induction seam welding.**

high-frequency upset welding (UW-HF). An upset welding process variation in which high-frequency welding current is supplied through electrodes into the workpieces. See Figure 51(A), 51(B), and 51(D). See also **high-frequency resistance welding** and **induction upset welding.**

high-low. A nonstandard term for **weld joint mismatch.**

high pulse current, *pulsed power welding.* The current during the high pulse time that produces the high heat level. See Figure 52.

high pulse time, *pulsed power welding.* The duration of the high pulse current. See Figure 52.

high vacuum electron beam welding (EBW-HV). An electron beam welding process variation in which welding is accomplished at a pressure of 10^{-4} to 10^{-1} pascals (approximately 10^{-6} to 10^{-3} torr).

hold time. The duration of force application at the point of welding after the last pulse ceases. See Figures 49 and 50.

hood. A nonstandard term for **welding helmet.**

horizontal fixed position. A nonstandard term for 5(G).

horizontal position. See **horizontal welding position.**

horizontal welding position, *fillet weld.* The welding position in which the weld is on the upper side of an approximately horizontal surface and against an approximately vertical surface. See Figures 16B, 18(B), 20(B), and 20(C).

horizontal welding position, *groove weld.* The welding position in which the weld face lies in an approximately vertical plane and the weld axis at the point of welding is approximately horizontal. See Figures 16A, 16(C), 17(B), and 19(B).

horizontal rolled position. A nonstandard term for 1G in pipe.

horn. An extension of the arm of a resistance welding machine that transmits the electrode force, usually conducts the welding current, and may support the workpiece.

horn spacing. A nonstandard term for **throat height.**

hot crack. A crack formed at temperatures near the completion of solidification.

hot isostatic pressure welding. A diffusion welding process variation that produces coalescence of metals by heating and applying hot inert gas under pressure.

hot pressure welding (HPW). A solid-state welding process that produces a weld with heat and application of pressure sufficient to produce macro deformation of the workpieces. See also **cold welding, diffusion welding,** and **forge welding.**

hot start current. A very brief current pulse at arc initiation to stabilize the arc quickly. See Figure 52.

hot wire welding. A variation of a fusion welding process in which a filler metal wire is resistance heated by current flowing through the wire as it is fed into the weld pool.

hydrogen brazing. A nonstandard term for any brazing process that takes place in a hydrogen atmosphere.

hydromatic welding. A nonstandard term for **pressure-controlled welding.**

I

impulse, *resistance welding.* A group of pulses occurring on a regular frequency separated only by an interpulse time. See Figures 48 and 49.

inclined position. A nonstandard term for **6G.**

inclined position, with restriction ring. A nonstandard term for **6GR.**

included angle. A nonstandard term for **groove angle.**

inclusion. Entrapped foreign solid material, such as slag, flux, tungsten, or oxide.

incomplete fusion. A weld discontinuity in which fusion did not occur between weld metal and fusion faces or adjoining weld beads. See Figure 29. See also **complete fusion.**

incomplete joint penetration. A joint root condition in a groove weld in which weld metal does not extend through the joint thickness. See Figure 26. See also **complete joint penetration, complete joint penetration weld, partial joint penetration weld,** and **joint penetration.**

indentation. In a spot, seam, or projection weld, the depression on the exterior surface of the workpieces.

indirect welding. A resistance welding secondary circuit variation in which the welding current flows through the workpieces in locations away from, as well as at, the welds for resistance spot, seam, or projection welding. See Figures 47(D), (E), (F), and (G).

induction brazing (IB). A brazing process that uses heat from the resistance of the workpieces to induced electric current.

induction seam welding (RSEW-I). A resistance seam welding process variation in which high-frequency welding current is induced in the workpieces. See also **high-frequency resistance welding** and **high-frequency seam welding.**

induction soldering (IS). A soldering process in which the heat required is obtained from the resistance of the workpieces to induced electric current.

induction upset welding (UW-I). An upset welding process variation in which high-frequency welding current is induced in the workpieces. See Figure 51(E). See also **high-frequency resistance welding** and **high-frequency upset welding.**

induction welding (IW). A welding process that produces coalescence of metals by the heat obtained from the resistance of the workpieces to the flow of induced high frequency welding current with or without the application of pressure. The effect of the high-frequency welding current is to concentrate the welding heat at the desired location. See Figure 51(E).

induction work coil. The inductor used when welding, brazing, or soldering with induction heating equipment. See Figure 51(E).

inert gas. A gas that normally does not combine chemically with materials. See also **protective atmosphere.**

inert gas metal arc welding. A nonstandard term for **gas metal arc welding.**

inert gas tungsten arc welding. A nonstandard term for **gas tungsten arc welding.**

inertia friction welding. A variation of friction welding in which the energy required to make the weld is supplied primarily by the stored rotational kinetic energy of the welding machine. See Figure 44.

infrared brazing (IRB). A brazing process that uses heat from infrared radiation.

infrared radiation. Electromagnetic energy with wave lengths from 770 to 12,000 nanometers.

infrared soldering (IRS). A soldering process in which the heat required is furnished by infrared radiation.

initial current. The current after starting, but before establishment of welding current. See Figure 52.

interface. See **braze interface, solder interface, thermal spray deposit interface,** and **weld interface.**

intergranular penetration. The penetration of a filler metal along the grain boundaries of a base metal.

intermediate flux. A soldering flux with a residue that generally does not attack the base metal. The original composition may be corrosive.

intermittent weld. A weld in which the continuity is broken by recurring unwelded spaces. See Figures 23(G) and (H).

interpass temperature, *thermal spraying.* In multipass thermal spraying, the temperature of the thermal spray area between thermal spray passes.

interpass temperature, *welding.* In a multipass weld, the temperature of the weld area between weld passes.

interpulse time, *resistance welding.* The time between successive pulses of current within the same impulse. See Figure 48.

interrupted spot welding. A nonstandard term for **multiple-impulse welding.**

ionic bond. A primary bond arising from the electrostatic attraction between two oppositely charged ions.

iron soldering (INS). A soldering process in which the heat required is obtained from a soldering iron.

J

J-groove weld. A type of groove weld. See Figures 8(E) and 9(D).

joint. The junction of members or the edges of members that are to be joined or have been joined. See Figures 1 and 2.

joint brazing procedure. The materials, detailed methods, and practices employed in the brazing of a particular joint.

joint buildup sequence. A nonstandard term for **cross-sectional sequence.**

joint clearance, *brazing and soldering.* The distance between the faying surfaces of a joint.

joint design. The shape, dimensions, and configuration of the joint. ||

joint efficiency. The ratio of strength of a joint to the strength of the base metal, expressed in percent.

joint filler. A metal plate inserted between the splice member and thinner joint member to accommodate joint members of dissimilar thickness in a spliced butt joint. See Figure 3(B). ||

joint geometry. The shape and dimensions of a joint in cross section prior to welding.

joint opening. A nonstandard term for **root opening.**

joint penetration. The distance the weld metal extends from the weld face into a joint, exclusive of weld reinforcement. See Figure 26. See also **groove weld size.** |

joint recognition. A function of an adaptive control that determines changes in the joint geometry during welding and directs the welding equipment to take appropriate action. See also **joint tracking** and **weld recognition.**

joint root. That portion of a joint to be welded where the members approach closest to each other. In cross section, the joint root may be either a point, a line, or an area. See Figure 4.

joint spacer. A metal part, such as a strip, bar, or ring, inserted in the joint root to serve as a backing and to maintain the root opening during welding. See Figure 24(F).

joint tracking. A function of an adaptive control that determines changes in joint location during welding and directs the welding machine to take appropriate action. See also **joint recognition** and **weld recognition.**

joint type. A weld joint classification based on five basic joint configurations such as a butt joint, corner joint, edge joint, lap joint, and T-joint. See Figures 1 and 2.

joint welding sequence. See **welding sequence.**

K

kerf. The width of the cut produced during a cutting process. See Figure 41.

keyhole welding. A technique in which a concentrated heat source penetrates partially or completely through a workpiece, forming a hole (keyhole) at the leading edge of the weld pool. As the heat source progresses, the molten metal fills in behind the hole to form the weld bead.

keying. A nonstandard term for **mechanical bond.**

knee. The supporting structure of the lower arm in a resistance welding machine.

knurling, *thermal spraying.* See **groove and rotary roughening, rotary roughening,** and **threading and knurling.**

L

lack of fusion. A nonstandard term for **incomplete fusion.**

lack of penetration. A nonstandard term for **incomplete joint penetration.**

lamellar tear. A subsurface terrace and step-like crack in the base metal with a basic orientation parallel to the wrought surface caused by tensile stresses in the through-thickness direction of the base metals weakened by the presence of small dispersed, planar shaped, nonmetallic inclusions parallel to the metal surface. See Figure 33(B).

lamination. A type of discontinuity with separation or weakness generally aligned parallel to the worked surface of a metal.

lance. See **oxygen lance** and **oxygen lance cutting.**

land. A nonstandard term for **root face.**

lap joint. A joint between two overlapping members in parallel planes. See Figures 1(D), 2(D), and 51(C).

laser. A device that produces a concentrated coherent light beam by stimulated electronic or molecular transitions to lower energy levels. Laser is an acronym for light amplification by stimulated emission of radiation.

laser beam air cutting (LBC-A). A laser beam cutting process variation that melts the workpiece and uses an air jet to remove molten and vaporized material.

laser beam braze welding (LBBW). A braze welding process variation that uses a laser beam as the heat source.

laser beam cutting (LBC). A thermal cutting process that severs metal by locally melting or vaporizing with the heat from a laser beam. The process is used with or without assist gas to aid the removal of molten and vaporized material. See also **laser beam air cutting, laser beam evaporative cutting, laser beam inert gas cutting,** and **laser beam oxygen cutting.**

laser beam cutting operator. See **thermal cutting operator.**

laser beam diameter. The diameter of a laser beam circular cross section at a specified location along the laser beam axis.

laser beam evaporative cutting (LBC-EV). A laser beam cutting process variation that vaporizes the workpiece, with or without an assist gas, typically inert gas, to aid the removal of vaporized material.

laser beam expander. A combination of optical elements that will increase the diameter of a laser beam.

laser beam inert gas cutting (LBC-IG). A laser beam cutting process variation that melts the workpiece and uses an inert assist gas to remove molten and vaporized material.

laser beam oxygen cutting (LBC-O). A laser beam cutting process variation that uses the heat from the chemical reaction between oxygen and the base metal at elevated temperatures. The necessary temperature is maintained with a laser beam.

laser beam splitter. An optical device that uses controlled reflection to produce two beams from a single incident beam.

laser beam welding (LBW). A welding process that produces coalescence with the heat from a laser beam impinging on the joint. The process is used without a shielding gas and without the application of pressure.

lasing gas. A gaseous lasing medium.

lasing medium. A material that emits coherent radiation by virtue of stimulated electronic or molecular transitions to lower energy.

|| **layer.** A stratum of weld metal consisting of one or more weld beads. See Figures 23(D) and 23(E).

layer level wound. A nonstandard term for **level wound.**

layer wound. A nonstandard term for **level wound.**

lead angle. A nonstandard term for **travel angle.**

lead burning. A nonstandard term when used for the welding of lead.

leg of a fillet weld. See **fillet weld leg.**

lens. See **filter lens.**

level wound. Spooled or coiled filler metal that has been wound in distinct layers such that adjacent turns touch. See also **random wound.**

lightly coated electrode. A filler metal electrode consisting of a metal wire with a light coating applied subsequent to the drawing operation, primarily for stabilizing the arc.

linear discontinuity. A discontinuity with a length that is substantially greater than its width.

linear indication. A test result in which a discontinuity in the material being tested is displayed as a linear or aligned array.

|| **liquation.** The partial melting of compositional heterogeneities such as banding or inclusion stringers in heated base metal or heat-affected zones.

liquidus. The lowest temperature at which a metal or an alloy is completely liquid.

local preheating. Preheating a specific portion of a structure.

local stress relief heat treatment. Stress relief heat treatment of a specific portion of a structure.

locked-up stress. A nonstandard term for **residual stress.**

longitudinal bend specimen. See **longitudinal weld test specimen.** ||

longitudinal crack. A crack with its major axis orientation approximately parallel to the weld axis. See Figure 33.

longitudinal sequence. The order in which the weld passes of a continuous weld are made with respect to its length. See also **backstep sequence, block sequence, cascade sequence, continuous sequence,** and **random sequence.**

longitudinal tension specimen. See **longitudinal weld test specimen.** ||

longitudinal weld test specimen. A weld test specimen with its major axis parallel to the weld axis. See also **transverse weld test specimen.** ||

low pulse current, *pulsed power welding.* The current during the low pulse time that produces the low heat level. See Figure 52.

low pulse time, *pulsed power welding.* The duration of the low current pulse. See Figure 52.

M

machine. A nonstandard term when used for **mechanized.**

machine welding. A nonstandard term when used for **mechanized welding.**

macroetch test. A test in which a specimen is prepared with a fine finish, etched, and examined under low magnification.

manifold. See **cylinder manifold.**

manual, *adj.* pertaining to the control of a process with the torch, gun, or electrode holder held and manipulated by hand. Accessory equipment, such as part motion devices and manually controlled material feeders may be used. See also **adaptive control, automatic, mechanized, robotic,** and **semiautomatic.** |

manual brazing. See **manual welding.**

manual soldering. See **manual welding.**

manual thermal cutting. See **manual welding.**

manual thermal spraying. See **manual welding.**

manual welding. Welding with the torch, gun, or electrode holder held and manipulated by hand. Accessory equipment, such as part motion devices and manually controlled filler material feeders may be used. Variations of this term are **manual brazing, manual soldering, manual thermal cutting,** and **manual thermal**

spraying. See Table 4. See also **adaptive control welding, automatic welding, mechanized welding, robotic welding,** and **semiautomatic welding.**

mash resistance seam welding. A nonstandard term for **mash seam welding.**

mash seam welding. A resistance seam welding process variation that makes a lap joint primarily by high-temperature plastic working and diffusion as opposed to melting and solidification. The joint thickness after welding is less than the original assembled thickness.

mask, *thermal spraying.* A device for protecting a substrate surface from the effects of blasting or adherence of a thermal spray deposit.

mechanical bond, *thermal spraying.* The adherence of a thermal spray deposit to a roughened surface by the mechanism of particle interlocking.

mechanized, *adj.* pertaining to the control of a process with equipment that requires manual adjustment of the equipment controls in response to visual observation of the operation, with the torch, gun, wire guide assembly, or electrode holder held by a mechanical device. See also **adaptive control, automatic, manual, robotic,** and **semiautomatic.**

mechanized brazing. See **mechanized welding.**

mechanized soldering. See **mechanized welding.**

mechanized thermal cutting. See **mechanized welding.**

mechanized thermal spraying. See **mechanized welding.**

mechanized welding. Welding with equipment that requires manual adjustment of the equipment controls in response to visual observation of the welding, with the torch, gun, or electrode holder held by a mechanical device. Variations of this term are **mechanized brazing, mechanized soldering, mechanized thermal cutting,** and **mechanized thermal spraying.** See Table 4. See also **adaptive control welding, automatic welding, manual welding, robotic welding,** and **semiautomatic welding.**

medium vacuum electron beam welding (EBW-MV). An electron beam welding process variation in which welding is accomplished at a pressure of 10^{-1} to 3×10^{3} pascals (approximately 10^{-3} to 25 torr).

meltback time. The time interval at the end of crater fill time to arc outage during which electrode feed is stopped. See Figure 53.

melting range. The temperature range between solidus and liquidus.

melting rate. The weight or length of electrode, wire, rod, powder melted in a unit of time.

melt-through. Visible root reinforcement produced in a joint welded from one side. See Figure 27.

metal. An opaque, lustrous, elemental chemical substance that is a good conductor of heat and electricity, usually malleable, ductile, and more dense than other elemental substances.

metal-bath dip brazing. A dip brazing process variation.

metal cored electrode. A composite tubular filler metal electrode consisting of a metal sheath and a core of various powdered materials, producing no more than slag islands on the face of a weld bead. External shielding may be required.

metal electrode. A filler or nonfiller metal electrode used in arc welding and cutting that consists of a metal wire or rod that has been manufactured by any method and that is either bare or covered.

metallic bond. The principal bond that holds metals together. It is a primary bond arising from the increased spatial extension of the valence electron wave functions when an aggregate of metal atoms is brought close together. See also **bonding force, covalent bond, ionic bond,** and **mechanical bond.**

metallizing. A nonstandard term when used for **thermal spraying** or the application of a metal coating.

metallurgical bond. A nonstandard term for **metallic bond.**

metal powder cutting (POC). An oxygen cutting process that uses heat from an oxyfuel gas flame, with iron or other metal powder to aid cutting.

microetch test. A test in which the specimen is prepared with a polished finish, etched, and examined under high magnification.

MIG welding. A nonstandard term for **gas metal arc welding** and **flux cored arc welding.**

mismatch. See **weld joint mismatch.**

mixed zone. The portion of the weld metal consisting of a mixture of base metal and filler metal. See also **unmixed zone.**

mixing chamber. That part of a welding or cutting torch in which a fuel gas and oxygen are mixed.

molding shoe. A nonstandard term for **backing shoe.**

molten weld pool. A nonstandard term for **weld pool.**

moving shoe. A backing shoe that slides along the joint during welding.

multiple-impulse welding. A resistance welding process variation in which welds are made by more than one impulse. See Figure 49.

multiport nozzle. A constricting nozzle of the plasma arc torch that contains two or more orifices located in a configuration to achieve some control over the arc shape.

|| **multiple welding position.** An orientation for a nonrotated circumferential joint requiring welding in more than one welding position. See **5F**, **5G**, **6F**, **6G**, and **6GR**.

N

|| narrow gap welding. A nonstandard term for **narrow groove welding.**

|| **narrow groove welding.** A variation of a welding process that uses multiple-pass welding with filler metal. The use of a small root opening, with either a square groove or a V groove and a small groove angle, yields a weld with a high ratio of depth to width.

neutral flame. An oxyfuel gas flame that has characteristics neither oxidizing nor reducing. See Figure 40. See also **carburizing flame**, **oxidizing flame**, and **reducing flame**.

neutral flux, *submerged arc welding.* A flux that will not cause a significant change in the weld metal composition when there is a large change in the arc voltage. See also **active flux**.

nonbutting member. A joint member that is free to move in any direction perpendicular to its thickness dimension. For example, both members of a lap joint, or one member of a T-joint or corner joint. See Figure 11. See also **butting member**.

nonconsumable electrode. An electrode that does not provide filler metal.

noncorrosive flux. A soldering flux that in either its original or residual form does not chemically attack the base metal. It usually is composed of rosin-base materials.

|| nondestructive evaluation. A nonstandard term for **nondestructive examination.**

|| **nondestructive examination (NDE).** The act of determining the suitability of some material or component for its intended purpose using techniques that do not affect its serviceability.

|| nondestructive inspection. A nonstandard term for **nondestructive examination**.

nondestructive testing. A nonstandard term for **nondestructive examination.** ||

nonsynchronous initiation. The closing of a resistance welding contractor without regard to the voltage wave form position.

nonsynchronous timing. A nonstandard term for **nonsynchronous initiation.**

nontransferred arc. An arc established between the electrode and the constricting nozzle of the plasma arc torch or thermal spraying gun. The workpiece is not in the electrical circuit. See also **transferred arc**.

nonvacuum electron beam welding (EBW-NV). An electron beam welding process variation in which welding is accomplished at atmospheric pressure.

nozzle. See **constricting nozzle** and **gas nozzle**.

nozzle, *arc spraying.* A device at the exit end of the gun that directs the atomizing air or other gas.

nozzle, *flame spraying.* A device at the exit end of the gun that directs and forms the flow shape of atomized spray particles and the accompanying air or other gases.

nozzle accumulation. Filler metal or surfacing material deposited on the inner surface and on the exit end of the nozzle.

nugget. The weld metal joining the workpieces in spot, seam, or projection welds.

nugget size. A nonstandard term when used for resistance spot weld size.

O

off time, *resistance welding.* The time during which the electrodes are off the workpieces. This term is generally used when the welding cycle is repetitive. See Figure 50.

1F, *pipe.* A welding test position designation for a circumferential fillet weld applied to a joint in pipe, with its axis approximately 45° from horizontal, in which the weld is made in the flat welding position by rotating the pipe about its axis. See Figure 20(A). ||

1F, *plate.* A welding test position designation for a linear fillet weld applied to a joint in which the weld is made in the flat welding position. See Figure 18(A). ||

1G, *pipe.* A welding test position designation for a circumferential groove weld applied to a joint in pipe, in which the weld is made in the flat welding position by rotating the pipe about its axis. See Figure 19(A). ||

1G, *plate.* A welding test position designation for a linear groove weld applied to a joint in which the weld is made in the flat welding position. See Figure 17(A).

open butt joint. A nonstandard term for a butt joint with an open root.

open circuit voltage. The voltage between the output terminals of the power source when no current is flowing to the torch or gun.

open groove. A nonstandard term for **open root joint**.

open joint. A nonstandard term for **open root joint**.

open root joint. An unwelded joint without backing or consumable insert.

orifice. See **constricting orifice**.

orifice gas. The gas that is directed into the plasma arc torch or thermal spraying gun to surround the electrode. It becomes ionized in the arc to form the arc plasma, and issues from the constricting orifice of the nozzle as a plasma jet. See Figure 35.

orifice throat length. The length of the constricting orifice in the plasma arc torch or thermal spraying gun.

oven soldering. A nonstandard term for **furnace soldering**.

overhang. A nonstandard term when used for **extension**.

overhead position. See **overhead welding position**.

overhead welding position. The welding position in which welding is performed from the underside of the joint. See Figures 16A, 16B, 16C, 17(D), 18(D), and 20(D).

overlap, *fusion welding.* The protrusion of weld metal beyond the weld toe or weld root. See Figures 32(C), and 32(D).

overlap, *resistance seam welding.* The portion of the preceding weld nugget remelted by the succeeding weld.

overlap. A nonstandard term when used for **incomplete fusion**.

overlaying. A nonstandard term when used for **surfacing**.

overspray, *thermal spraying.* The portion of the thermal spray deposit that is not deposited on the workpiece.

oxidizing flame. An oxyfuel gas flame in which there is an excess of oxygen, resulting in an oxygen-rich zone extending around and beyond the cone. See Figure 40. See also **carburizing flame**, **neutral flame**, and **reducing flame**.

oxyacetylene cutting (OFC-A). An oxyfuel gas cutting process variation that uses acetylene as the fuel gas.

oxyacetylene welding (OAW). An oxyfuel gas welding process that uses acetylene as the fuel gas. The process is used without the application of pressure. See Figure 40.

oxyfuel gas cutting (OFC). A group of oxygen cutting processes that uses heat from an oxyfuel gas flame. See also **oxyacetylene cutting, oxyhydrogen cutting, oxynatural gas cutting,** and **oxypropane cutting.**

oxyfuel gas cutting torch. A device used for directing the preheating flame produced by the controlled combustion of fuel gases and to direct and control the cutting oxygen.

oxyfuel gas spraying. A nonstandard term for **flame spraying.**

oxyfuel gas welding (OFW). A group of welding processes that produces coalescence of workpieces by heating them with an oxyfuel gas flame. The processes are used with or without the application of pressure and with or without filler metal.

oxyfuel gas welding torch. A device used in oxyfuel gas welding, torch brazing, and torch soldering for directing the heating flame produced by the controlled combustion of fuel gases.

oxygas cutting. A nonstandard term for **oxyfuel gas cutting.**

oxygen arc cutting (AOC). An oxygen cutting process that uses an arc between the workpiece and a consumable tubular electrode, through which oxygen is directed to the workpiece.

oxygen cutter. See **thermal cutter.**

oxygen cutting (OC). A group of thermal cutting processes that severs or removes metal by means of the chemical reaction between oxygen and the base metal at elevated temperature. The necessary temperature is maintained by the heat from an arc, an oxyfuel gas flame, or other source.

oxygen cutting operator. See **thermal cutting operator.**

oxygen gouging. Thermal gouging that uses an oxygen cutting process variation to form a bevel or groove.

oxygen grooving. A nonstandard term for **oxygen gouging.**

oxygen lance. A length of pipe used to convey oxygen to the point of cutting in oxygen lance cutting.

oxygen lance cutting (LOC). An oxygen cutting process that uses oxygen supplied through a consumable lance. The preheat to start the cutting is obtained by other means.

oxygen lancing. A nonstandard term for **oxygen lance cutting.**

oxyhydrogen cutting (OFC-H). An oxyfuel gas cutting process variation that uses hydrogen as the fuel gas.

oxyhydrogen welding (OHW). An oxyfuel gas welding process that uses hydrogen as the fuel gas. The process is used without the application of pressure.

oxynatural gas cutting (OFC-N). An oxyfuel gas cutting process variation that uses natural gas as the fuel gas.

oxypropane cutting (OFC-P). An oxyfuel gas cutting process variation that uses propane as the fuel gas.

P

parallel gap welding. A nonstandard term for series welding with closely spaced electrodes.

parallel welding. A resistance welding secondary circuit variation in which the secondary current is divided and conducted through the workpieces and electrodes in parallel electrical paths to simultaneously form multiple resistance spot, seam, or projection welds. See Figures 46(A) and (B).

parent metal. A nonstandard term for **base metal** or **substrate.**

|| **partial joint penetration weld.** A joint root condition in a groove weld in which incomplete joint penetration exists. See Figure 26. See also **complete joint penetration, complete joint penetration weld, incomplete joint penetration,** and **joint penetration.**

pass. See **thermal spraying pass** and **weld pass.**

pass sequence. See **weld pass sequence.**

| **paste brazing filler metal.** A mixture of finely divided brazing filler metal with a flux or neutral carrier.

|| **paste solder.** A mixture of finely divided solder with a flux or neutral carrier.

peel test. A destructive method of testing that mechanically separates a lap joint by peeling.

peening. The mechanical working of metals using impact blows.

| penetration. A nonstandard term when used for **depth of fusion, joint penetration,** or **root penetration.**

percent ferrite. A nonstandard term when used for **Ferrite Number.**

percussion welding (PEW). A welding process that produces coalescence with an arc resulting from a rapid discharge of electrical energy. Pressure is applied percussively during or immediately following the electrical discharge.

pilot arc. A low current arc between the electrode and the constricting nozzle of the plasma arc torch to ionize the gas and facilitate the start of the welding arc.

plasma. See **arc plasma.**

plasma arc cutting (PAC). An arc cutting process that uses a constricted arc and removes the molten metal with a high-velocity jet of ionized gas issuing from the constricting orifice.

plasma arc cutting torch. A device used to transfer current to a fixed cutting electrode, position the electrode, and direct the flow of shielding gas and orifice gas. See Figure 35. ||

plasma arc welding (PAW). An arc welding process that uses a constricted arc between a nonconsumable electrode and the weld pool (transferred arc) or between the electrode and the constricting nozzle (nontransferred arc). Shielding is obtained from the ionized gas issuing from the torch, which may be supplemented by an auxiliary source of shielding gas. The process is used without the application of pressure. See also **hot wire welding.**

plasma arc welding torch. A device used to transfer current to a fixed welding electrode, position the electrode, and direct the flow of shielding gas and orifice gas. See Figure 35. ||

plasma sprayer. See **thermal sprayer.**

plasma spraying (PSP). A thermal spraying process in which a nontransferred arc of the gun is used to create an arc plasma for melting and propelling the surfacing material to the substrate.

plasma spraying operator. See **thermal spraying operator.**

platen, *resistance welding.* A member with a substantially flat surface to which dies, fixtures, backups, or electrode holders are attached, and that transmits the electrode force or upset force. One platen usually is fixed and the other moveable. |

platen spacing. The distance between adjacent surfaces of the platens in a resistance welding machine.

plenum. See **plenum chamber.**

plenum chamber. The space between the electrode and the inside wall of the constricting nozzle of the plasma arc torch or thermal spraying gun. See Figure 35. |

plug weld. A weld made in a circular hole in one member of a joint fusing that member to another member. A fillet-welded hole is not to be construed as conforming to this definition. See Figure 15(E).

|| **plug weld size.** The diameter of the weld metal in the plane of the faying surfaces.

poke welding. A nonstandard term for **push welding.**

polarity. See **direct current electrode negative** and **direct current electrode positive.**

porosity. Cavity-type discontinuities formed by gas entrapment during solidification or in a thermal spray deposit.

position. See **welding position.**

position of welding. See **welding position.**

postflow time. The time interval from current shut off to either shielding gas or cooling water shut off. See Figures 52 and 53.

postheating. The application of heat to an assembly after welding, brazing, soldering, thermal spraying, or thermal cutting.

postweld interval, *resistance welding.* The total elapsed time from the end of the weld interval to the end of hold time. See Figure 49.

powder alloy. A nonstandard term for **alloy powder.**

powder blend. A heterogeneous mixture of two or more alloy, metal, or nonmetal powders. See also **alloy powder.**

powder composite. Two or more different materials combined to form a single particle, formed by either chemical coating or mechanical agglomeration.

powder cutting. A nonstandard term for **flux cutting** and **metal powder cutting.**

powder feeder. A device for supplying powdered surfacing material to a thermal spraying gun or cutting torch.

powder feed gas. A nonstandard term for **carrier gas.**

powder feed rate. The quantity of powder fed to a thermal spraying gun or a cutting torch per unit of time.

powder flame spraying. A flame spraying process variation in which the surfacing material is in powder form. See also **flame spraying.**

power source. An apparatus for supplying current and voltage suitable for welding, thermal cutting, or thermal spraying.

precoating. Coating the base metal in the joint by dipping, electroplating, or other applicable means prior to soldering or brazing.

preflow time. The time interval between start of shielding gas flow and arc starting. See Figures 52 and 53.

preform. Brazing or soldering filler metal fabricated in a shape or form for a specific application.

preheat. The heat applied to the base metal or substrate to attain and maintain preheat temperature.

preheat current, *resistance welding.* An impulse or series of impulses that occur prior to and are separated from the welding current. See Figure 49.

preheat temperature, *brazing and soldering.* The temperature of the base metal in the volume surrounding the point of brazing or soldering immediately before brazing or soldering is started. ||

preheat temperature, *thermal cutting.* The temperature of the base metal in the volume surrounding the point of thermal cutting immediately before thermal cutting is started. ||

preheat temperature, *thermal spraying.* The temperature of the substrate in the volume surrounding the point of thermal spraying immediately before thermal spraying is started. In a multipass thermal spraying, it is also the temperature immediately before the second and subsequent passes are started. ||

preheat temperature, *welding.* The temperature of the base metal in the volume surrounding the point of welding immediately before welding is started. In a multipass weld, it is also the temperature immediately before the second and subsequent passes are started. ||

preheat time, *resistance welding.* The duration of preheat current flow during the preweld interval. See Figure 49.

prequalified welding procedure specification. A welding procedure specification that complies with the stipulated conditions of a particular welding code or specification and is therefore acceptable for use under that code or specification without a requirement for qualification testing. ||

pressure-controlled welding. A resistance welding process variation in which a number of spot or projection welds are made with several electrodes functioning progressively under the control of a pressure-sequencing device.

pressure gas welding (PGW). An oxyfuel gas welding process that produces a weld simultaneously over the entire faying surfaces. The process is used with the application of pressure and without filler metal.

pressure welding. A nonstandard term for **solid-state welding, hot pressure welding, forge welding, diffusion welding, pressure gas welding,** and **cold welding.**

pretinning. A nonstandard term for **precoating.**

preweld interval, *resistance welding.* The elapsed time between the initiation of the squeeze time and the beginning of the weld time or weld interval time. See Figure 49.

procedure. The detailed elements of a process or method used to produce a specific result.

procedure qualification. The demonstration that welds made by a specific procedure can meet prescribed standards.

procedure qualification record (PQR). See **brazing procedure qualification record** and **welding procedure qualification record.**

process. A grouping of basic operational elements used in welding, thermal cutting, or thermal spraying. See also the Master Chart of Welding and Allied Processes.

progressive block sequence. A block sequence in which successive blocks are completed progressively along the weld, either from one end to the other or from an intermediate location of the weld toward either end.

projection welding (PW). A resistance welding process that produces a weld by the heat obtained from the resistance to the flow of the welding current. The resulting welds are localized at predetermined points by projections, embossments, or intersections.

|| **projection weld size.** The diameter of the weld metal in the plane of the faying surfaces. See Figure 25(F).

protective atmosphere. A gas or vacuum envelope surrounding the workpieces, used to prevent or reduce the formation of oxides and other detrimental surface substances, and to facilitate their removal.

puddle. A nonstandard term when used for **weld pool.**

puddle weld. A nonstandard term for an **arc spot weld** or **plug weld.**

pull gun technique. A nonstandard term for **backhand welding.**

pulsation welding. A nonstandard term for **multiple-impulse welding.**

pulse, *resistance welding.* A current of controlled duration of either polarity through the welding circuit. See Figures 48 and 49.

pulsed gas metal arc welding (GMAW-P). A gas metal arc welding process variation in which the current is pulsed. See also **pulsed power welding.**

pulsed gas tungsten arc welding (GTAW-P). A gas tungsten arc welding process variation in which the current is pulsed. See also **pulsed power welding.**

pulsed laser. A laser whose output is controlled to produce a pulse whose duration is 25 milliseconds or less.

pulsed power welding. An arc welding process variation in which the power is cyclically programmed to pulse so that effective but short duration values of power can be utilized. Such short duration values are significantly different from the average value of power. Equivalent terms are pulsed voltage or pulsed current welding. See also **pulsed spray welding.**

pulsed spray welding. An arc welding process variation in which the current is pulsed to utilize the advantages of the spray mode of metal transfer at average currents equal to or less than the globular to spray transition current.

pulse start delay time. The time interval from current initiation to the beginning of current pulsation. See Figure 52.

pulse time, *resistance welding.* The duration of a pulse. See Figure 48.

push angle. The travel angle when the electrode is pointing in the direction of weld progression. This angle can also be used to partially define the position of guns, torches, rods, and beams. See Figure 21. See also **drag angle, forehand welding, travel angle,** and **work angle.**

push welding. A resistance welding process variation in which spot or projection welds are made by manually applying force to one electrode and using the workpiece or a support as the other electrode.

Q

qualification. See **welder performance qualification** and **procedure qualification.**

quench time, *resistance welding.* The time from the end of the weld, weld interval, or downslope time to the beginning of the temper time, during which no current flows through the workpieces and the weld is rapidly cooled by the electrodes. See Figure 49.

R

random intermittent welds. Intermittent welds on one or both sides of a joint in which the weld increments are made without regard to spacing.

random sequence. A longitudinal sequence in which the weld bead increments are made at random.

random wound. Spooled or coiled filler metal that has not been wound in distinct layers. See also **level wound.**

rate of deposition. See **deposition rate.**

rate of flame propagation. See **flame propagation rate.**

reaction flux, *soldering.* A flux composition in which one or more of the ingredients reacts with a base metal upon heating to deposit one or more metals.

reaction soldering. A soldering process variation in which a reaction flux is used.

reaction stress. A stress that cannot exist in a member if the member is isolated as a free body without connection to other parts of the structure.

reactor. A device used in arc welding circuits to minimize irregularities in the flow of the welding current.

reduced section tension test. A test in which a transverse section of the weld is located in the center of the reduced section of the specimen.

reducing atmosphere. A chemically active protective atmosphere that will reduce metal oxides to their metallic state at elevated temperature.

reducing flame. An oxyfuel gas flame with an excess of fuel gas. See also **carburizing flame, neutral flame, oxidizing flame**, and **reducing atmosphere**.

reflowing. A nonstandard term when used for **flow brightening**.

reflow soldering. A nonstandard term for soldering with preplaced filler metal.

reinforcement of weld. See **weld reinforcement**.

residual stress. Stress present in a joint member or material that is free of external forces or thermal gradients.

resistance brazing (RB). A brazing process that uses heat from the resistance to electric current flow in a circuit of which the workpieces are a part.

resistance butt welding. A nonstandard term for **upset welding** and **flash welding**.

resistance seam welding (RSEW). A resistance welding process that produces a weld at the faying surfaces of overlapped parts progressively along a length of a joint. The weld may be made with overlapping weld nuggets, a continuous weld nugget, or by forging the joint as it is heated to the welding temperature by resistance to the flow of the welding current. See Figures 14(D), 30(D), and 51. See also **high frequency seam welding** and **induction seam welding.**

resistance seam weld size. See **seam weld size.** |

resistance soldering (RS). A soldering process that uses heat from the resistance to electric current flow in a circuit of which the workpieces are a part.

resistance spot welding (RSW). A resistance welding process that produces a weld at the faying surfaces of a joint by the heat obtained from resistance to the flow of welding current through the workpieces from electrodes that serve to concentrate the welding current and pressure at the weld area. See Figures 14(E), 30(D), and 46–50.

resistance spot weld size. See **spot weld size.** ||

resistance welding (RW). A group of welding processes that produces coalescence of the faying surfaces with the heat obtained from resistance of the workpieces to the flow of the welding current in a circuit of which the workpieces are a part, and by the application of pressure. ||

resistance welding control. The device, usually electronic, that determines the welding sequence and timing with regard to the welding current pattern, electrode or platen force or movement, and other operational conditions of a resistance welding machine.

resistance welding current. The current in the welding circuit during the making of a weld, but excluding preweld or postweld current. See Figures 49 and 50.

resistance welding die. A resistance welding electrode usually shaped to the workpiece contour to clamp the workpieces and to conduct the welding current.

resistance welding downslope time. The time during which the welding current is continuously decreased. See Figure 49.

resistance welding electrode. The part of a resistance welding machine through which the welding current and, in most cases, force are applied directly to the workpiece. The electrode may be in the form of a rotating wheel, rotating roll, bar, cylinder, plate, clamp, chuck, or modification thereof.

resistance welding gun. A manipulatable device to transfer current and provide electrode force to the weld area (usually in reference to a portable gun).

resistance welding upslope time. The time during which the welding current continuously increases from the beginning of the welding current. See Figure 49.

resistance welding weld time. The duration of welding current flow through the workpieces in making a weld by single-impulse welding or flash welding. See Figure 50. See also **weld interval.**

retaining shoe. A nonstandard term for **backing shoe.**

reverse polarity. A nonstandard term for **direct current electrode positive.**

robotic, *adj.* pertaining to process control by robotic equipment. See also **adaptive control, automatic, manual, mechanized,** and semiautomatic.

robotic brazing. See **robotic welding.**

robotic soldering. See **robotic welding.**

robotic thermal cutting. See **robotic welding.**

robotic thermal spraying. See **robotic welding.**

robotic welding. Welding that is performed and controlled by robotic equipment. Variations of this term are **robotic brazing, robotic soldering, robotic thermal cutting,** and **robotic thermal spraying.** See Table 4. See also **adaptive control welding, automatic welding, manual welding, mechanized welding,** and **semiautomatic welding.**

rollover. A nonstandard term when used for **overlap.**

roll spot welding. A resistance welding process variation that makes intermittent spot welds using one or more rotating circular electrodes. The rotation of the electrodes may or may not be stopped during the making of a weld.

roll welding (ROW). A solid-state welding process that produces a weld by the application of heat and sufficient pressure with rolls to cause deformation at the faying surfaces. See also **forge welding.**

root. A nonstandard term when used for **joint root** and **weld root.**

root bead. A weld bead that extends into or includes part or all of the joint root.

root bend test. A test in which the weld root is on the convex surface of a specified bend radius.

root crack. See Figure 33.

root edge. A root face of zero width. See Figure 5.

root face. That portion of the groove face within the joint root. See Figure 5.

root gap. A nonstandard term for root opening.

root of joint. See **joint root.**

root of weld. See **weld root.**

root opening. A separation at the joint root between the workpieces. See Figures 6 and 25(D).

root penetration. The distance the weld metal extends into the joint root. See Figure 26.

root radius. A nonstandard term for **groove radius.**

root reinforcement. Weld reinforcement opposite the side from which welding was done. See Figure 24A. See also **face reinforcement.**

root surface. The exposed surface of a weld opposite the side from which welding was done. See Figures 24(B), 27(E), and 27(F).

root surface crack. See Figure 33.

root surface underfill. See **underfill.** See Figure 32(E).

rotary roughening, *thermal spraying.* A method of surface roughening in which a revolving tool is pressed against the surface being prepared, while either the work or the tool, or both, move. See Figure 43(D).

rough threading, *thermal spraying.* A method of surface roughening that consists of cutting threads with the sides and tops of the threads jagged and torn.

round edge shape. A type of edge shape. See Figure 7(G).

runoff weld tab. Additional material that extends beyond the end of the joint, on which the weld is terminated.

S

salt-bath dip brazing. A dip brazing process variation.

scarf. A nonstandard term for **bevel.**

scarf groove. A weld groove in a butt joint consisting of members having single-bevel edge shapes. The groove faces are parallel. See Figure 13(B).

scarf joint. A nonstandard term for scarf groove.

seal-bonding material, *thermal spraying.* A material that partially forms, in the as-sprayed condition, a metallic bond with the substrate.

seal coat, *thermal spraying.* Material applied to infiltrate and close the pores of a thermal spray deposit.

seal weld. Any weld designed primarily to provide a specific degree of tightness against leakage.

seam. A nonstandard term when used for a welded, brazed, or soldered joint.

seam weld. A continuous weld made between or upon overlapping members, in which coalescence may start and occur on the faying surfaces, or may have proceeded from the outer surface of one member. The continuous weld may consist of a single weld bead or a series of overlapping spot welds. See Figures 14 and 51(C). See also **arc seam weld** and **resistance seam welding**.

|| **seam weld size.** The width of the weld metal in the plane of the faying surfaces. See Figures 25(F) and 25(G).

secondary circuit. That portion of a welding machine that conducts the secondary current between the secondary terminals of the welding transformer and the electrodes, or electrode and workpiece.

secondary current path, *resistance welding.* The electrical path through which the welding current passes.

selective block sequence. A block sequence in which successive blocks are completed in an order selected to control residual stresses and distortion.

self-fluxing alloy, *thermal spraying.* A surfacing material that wets the substrate and coalesces when heated to its melting point, with no flux other than the boron and silicon contained in the alloy.

|| **self-shielded flux cored arc welding (FCAW-S).** A flux cored arc welding process variation in which shielding gas is obtained exclusively from the flux within the electrode.

semiautomatic, *adj.* pertaining to the manual control of a process with equipment that automatically controls one or more of the process conditions. See also **adaptive control, automatic, manual, mechanized,** and **robotic.**

semiautomatic brazing. See **semiautomatic welding.**

semiautomatic soldering. See **semiautomatic welding.**

semiautomatic thermal cutting. See **semiautomatic welding.**

semiautomatic thermal spraying. See **semiautomatic welding.**

semiautomatic welding. Manual welding with equipment that automatically controls one or more of the welding conditions. See Table 4. Variations of this term are **semiautomatic brazing, semiautomatic soldering, semiautomatic thermal cutting,** and **semiautomatic thermal spraying.** See also **adaptive control welding, automatic welding, manual welding, mechanized welding,** and **robotic welding.**

semiblind joint. A joint in which one extremity of the joint is not visible.

sequence time. A nonstandard term when used for **welding cycle.**

series submerged arc welding (SAW-S). A submerged arc welding process variation in which the arc is established between two consumable electrodes that meet just above the surface of the workpieces, which are not part of the welding current circuit.

series welding. A resistance welding secondary circuit variation in which the secondary current is conducted through the workpieces and electrodes or wheels in a series electrical path to simultaneously form multiple resistance spot, seam, or projection welds. See Figures 46(C) and 46(D).

setback. See **contact tube setback** and **electrode setback.** ||

set down. A nonstandard term when used for **upset.**

shadow mask, *thermal spraying.* A device that partially shields an area of the workpiece, producing a feathered edge of the thermal spray deposit.

sheet separation, *resistance welding.* The distance between the faying surfaces, adjacent to the weld, after a spot, seam, or projection weld has been made.

shielded carbon arc welding (CAW-S). A carbon arc welding process variation that uses shielding from the combustion of solid material fed into the arc, or from a blanket of flux on the workpieces, or both.

shielded metal arc cutting (SMAC). An arc cutting process that uses a covered electrode.

shielded metal arc welding (SMAW). An arc welding process with an arc between a covered electrode and the weld pool. The process is used with shielding from the decomposition of the electrode covering, without the application of pressure, and with filler metal from the electrode. See also **firecracker welding.**

shielding gas. Protective gas used to prevent or reduce atmospheric contamination. See also **protective atmosphere.**

short arc. A nonstandard term when used for **short circuit gas metal arc welding** or **short circuiting transfer.**

short circuit gas metal arc welding (GMAW-S). A gas metal arc welding process variation in which the consumable electrode is deposited during repeated short circuits.

short circuiting arc welding. A nonstandard term for **short circuit gas metal arc welding.**

short circuiting transfer, *arc welding.* Metal transfer in which molten metal from a consumable electrode is deposited during repeated short circuits. See Figure 39(B). See also **globular transfer** and **spray transfer.**

shoulder. A nonstandard term when used for **root face.**

shrinkage stress. A nonstandard term when used for **residual stress.**

shrinkage void. A cavity-type discontinuity normally formed by shrinkage during solidification.

sidewall. A nonstandard term when used for **groove face.**

sieve analysis. A method of determining particle size distribution, usually expressed as the weight percentage retained upon each of a series of standard screens of decreasing mesh size.

side bend test. A test in which the side of a transverse section of the weld is on the convex surface of a specified bend radius.

silver alloy brazing. A nonstandard term for brazing with a silver-base filler metal.

silver soldering. A nonstandard term for brazing with a silver-base filler metal.

single-bevel edge shape. A type of edge shape. See Figure 7(B).

single-bevel-groove weld. A type of groove weld. See Figure 8(B).

single-flare-bevel-groove weld. A weld in a groove formed by a member with a curved surface in contact with a planar member. See Figure 8(G).

single-flare-V-groove weld. A weld in a groove formed by two members with curved surfaces. See Figure 8(H).

single-groove weld, *fusion welding*. A groove weld that is made from one side only. See Figure 8.

single impulse welding. A resistance welding process variation in which spot, projection, or upset welds are made with a single pulse. See Figure 50.

single-J edge shape. A type of edge shape. See Figure 7(D).

single-J-groove weld. A type of groove weld. See Figure 8(E).

single-port nozzle. A constricting nozzle of the plasma arc torch that contains one orifice, located below and concentric with the electrode.

|| **single-spliced butt joint.** See **spliced joint**. See Figure 3(A).

|| **single-spliced joint.** See **spliced joint**. See Figure 3(A).

single-square-groove weld. A type of groove weld. See Figure 8(A).

single-U-groove weld. A type of groove weld. See Figure 8(F).

single-V-groove weld. A type of groove weld. See Figures 8(C) and 8(D).

single-welded joint, *fusion welding*. A joint that is welded from one side only. See Figure 8. |

6F. A welding test position designation for a circumferential fillet weld applied to a joint in pipe, with its axis approximately 45° from horizontal, in which the weld is made in flat, vertical, and overhead welding positions. The pipe remains fixed until welding is complete. See Figure 20(F). ||

6G. A welding test position designation for a circumferential groove weld applied to a joint in pipe, with its axis approximately 45° from horizontal, in which the weld is made in the flat, vertical, and overhead welding positions. The pipe remains fixed until welding is complete. See Figure 19(D). ||

6GR. A welding test position designation for a circumferential groove weld applied to a joint in pipe, with its axis approximately 45° from horizontal, in which the weld is made in the flat, vertical, and overhead welding positions. A restriction ring is added, adjacent to the joint, to restrict access to the weld. The pipe remains fixed until welding is complete. See Figure 19(E). ||

size of weld. See **weld size.**

skip weld. A nonstandard term for **intermittent weld.**

skull. The unmelted residue from a liquated filler metal.

slag. A nonmetallic product resulting from the mutual dissolution of flux and nonmetallic impurities in some welding and brazing processes.

slot weld. A weld made in an elongated hole in one member of a joint fusing that member to another member. The hole may be open at one end. A fillet welded slot is not to be construed as conforming to this definition. See Figure 15(D).

slot weld size. The width and length of the weld metal in the plane of the faying surfaces.

slugging. The unauthorized addition of metal, such as a length of rod, to a joint before welding or between passes, often resulting in a weld with incomplete fusion.

smoothing bead. A weld bead made to correct an undesirable weld profile, but not to enhance weld appearance. See also **cosmetic pass.**

smoothing pass. A weld pass that results in a smoothing bead.

soft solder. A nonstandard term for **solder.**

solder. The metal or alloy used as a filler metal in soldering, which has a liquidus not exceeding 450°C (840°F) and below the solidus of the base metal.

solderability. The capacity of a material to be soldered under the imposed fabrication conditions into a specific, suitably designed structure and to perform satisfactorily in the intended service.

soldering (S). A group of welding processes that produces coalescence of materials by heating them to the soldering temperature and by using a filler metal having a liquidus not exceeding 450°C (840°F) and below the solidus of the base metals. The filler metal is distributed between closely fitted faying surfaces of the joint by capillary action.

soldering blowpipe. A device used to obtain a small, accurately directed flame for fine work. A portion of any flame is blown to the desired location by the blowpipe, which is usually mouth operated.

soldering gun. An electrical soldering iron with a pistol grip and a quick heating, relatively small bit.

soldering iron. A soldering tool having an internally or externally heated metal bit usually made of copper.

solder interface. The interface between solder metal and base metal in a soldered joint. See Figure 31(A).

solder metal. That portion of a soldered joint that has been melted during soldering.

solid-state welding (SSW). A group of welding processes that produces coalescence by the application of pressure at a welding temperature below the melting temperatures of the base metal and the filler metal.

solidus. The highest temperature at which a metal or an alloy is completely solid.

spacer. See **joint spacer.**

spacer strip. A metal strip or bar prepared for a groove weld and inserted in the joint root to serve as a backing and to maintain the root opening during welding. It can also bridge an exceptionally wide root opening due to poor fit. See Figure 24(F).

spark shield. A nonstandard term for a safety shield.

spatter. The metal particles expelled during fusion welding that do not form a part of the weld.

spatter loss. Metal lost due to spatter.

spiking, *electron beam welding and laser beam welding.* A condition where the joint penetration is nonuniform and changes abruptly over the length of the weld.

spit. A nonstandard term when used for **flash** and **expulsion.**

splice. A nonstandard term when used for a welded, brazed, or soldered joint.

spliced butt joint. See **spliced joint.** See Figures 3(A) and (B).

spliced joint. A joint in which an additional workpiece spans the joint and is welded to each joint member. See Figures 3(A) and (B). See also **splice member.**

splice member. The workpiece that spans the joint in a spliced joint. See Figures 3(A) and (B).

split pipe backing. A pipe segment used as a backing for welding butt joints in round bars. See Figure 12.

spool. A filler metal package consisting of a continuous length of welding wire in coil form wound on a cylinder (called a barrel), which is flanged at both ends. The flange contains a spindle hole of smaller diameter than the inside diameter of the barrel. See Figure 42(A).

spot weld. A weld made between or upon overlapping members in which coalescence may start and occur on the faying surfaces or may proceed from the outer surface of one member. The weld cross section (plan view) is approximately circular. See Figure 14(E), 14(F) and 30(D). See also **arc spot weld** and **resistance spot welding.**

spot weld size. The diameter of the weld metal in the plane of the faying surfaces. See Figures 25(F), 25(G) and 30(D).

spray arc. A nonstandard term for **spray transfer.**

spray deposit. See **thermal spray deposit.**

spray deposit density ratio. See **thermal spray deposit density ratio.**

sprayer. See **thermal sprayer.**

spraying booth. An exhaust booth where thermal spraying is performed.

spraying operator. See **thermal spraying operator.**

spraying rate, *thermal spraying.* The rate at which surfacing material passes through the gun.

spraying sequence, *thermal spraying.* The order in which layers of materials are applied, such as overlapped, superimposed, or at various angles.

spray tab, *thermal spraying.* A small piece of additional material that is thermally sprayed concurrently with the workpiece, and used to evaluate the quality of the thermal spray deposit.

spray transfer, *arc welding.* Metal transfer in which molten metal from a consumable electrode is propelled axially across the arc in small droplets. See Figure 39(C). See also **globular transfer** and **short circuiting transfer.**

square edge shape. A type of edge shape. See Figure 7(A).

square-groove weld. A type of groove weld. See Figures 8(A) and 9(A).

squeeze time, *resistance welding.* The time between the initiation of the welding cycle and first application of current in spot, seam, or projection and some types of upset welds. See Figures 49 and 50.

stack cutting. Thermal cutting of stacked metal plates arranged so that all the plates are severed by a single cut.

staggered intermittent weld. An intermittent weld on both sides of a joint in which the weld increments on one side are alternated with respect to those on the other side. See Figure 23(H).

standoff distance. The distance between a nozzle and the workpiece. See Figures 35, 36, and 38.

start current. The current value during start time interval. See Figure 53.

starting weld tab. Additional material that extends beyond the beginning of the joint, on which the weld is started.

start time. The time interval prior to weld time during which arc voltage and current reach a preset value greater or less than welding values. See Figure 53.

static electrode force. The force exerted by electrodes on the workpieces in making spot, seam, or projection welds by resistance welding under welding conditions, but with no current flowing and no movement in the welding machine.

stationary shoe. A backing shoe that remains in a fixed position during welding.

stepback sequence. A nonstandard term for **backstep sequence.**

step brazing. The brazing of successive joints on a given part with filler metals of successively lower brazing temperatures so as to accomplish the joining without disturbing the joints previously brazed.

step soldering. The soldering of successive joints on a given part with solders of successively lower soldering temperature so as to accomplish the joining without disturbing the joints previously soldered.

stick electrode. A nonstandard term for **covered electrode.**

stick electrode welding. A nonstandard term for **shielded metal arc welding.**

stickout, *gas tungsten arc welding.* The length of the tungsten electrode extending beyond the end of the gas nozzle. See Figure 36.

stickout, *gas metal arc welding and gas shielded flux cored arc welding.* The length of unmelted electrode extending beyond the end of the gas nozzle. See Figure 38.

stitch weld. A nonstandard term for **intermittent weld.**

stopoff. A material used on the surfaces adjacent to the joint to limit the spread of soldering or brazing filler metal.

stored energy welding. A resistance welding process variation in which welds are made with electrical energy accumulated electrostatically, electromagnetically, or electrochemically at a relatively low rate and made available at the required welding rate.

straight polarity. A nonstandard term for **direct current electrode negative.**

stranded electrode. A composite filler metal electrode consisting of stranded wires that may mechanically enclose materials to improve properties, stabilize the arc, or provide shielding.

stress-corrosion cracking. Failure of metals by cracking under combined action of corrosion and stress, residual or applied. In brazing, the term applies to the cracking of stressed base metal due to the presence of a liquid filler metal.

stress-relief cracking. Intergranular cracking in the heat-affected zone or weld metal as a result of the combined action of residual stresses and postweld exposure to an elevated temperature.

stress-relief heat treatment. Uniform heating of a structure or a portion thereof to a sufficient temperature to relieve the major portion of the residual stresses, followed by uniform cooling.

strike. See **arc strike.**

stringer bead. A type of weld bead made without appreciable weaving motion. See Figure 22(A). See also **weave bead.**

strongback. A device attached to both sides of a weld joint to maintain alignment of the members during welding.

stub. The short length of filler metal electrode, welding rod, or brazing rod that remains after its use for welding or brazing.

stud arc welding. A nonstandard term for **arc stud welding.**

stud welding. A general term for joining a metal stud or similar part to a workpiece. Welding may be accomplished by arc, resistance, friction, or other process with or without external gas shielding. See also **arc stud welding.**

submerged arc welding (SAW). An arc welding process that uses an arc or arcs between a bare metal electrode or electrodes and the weld pool. The arc and molten metal are shielded by a blanket of granular flux on the workpieces. The process is used without pressure and with filler metal from the electrode and sometimes from a supplemental source (welding rod, flux, or metal granules). See also **hot wire welding** and **series submerged arc welding.**

substrate. Any material to which a thermal spray deposit is applied.

suck-back. A nonstandard term for **concave root surface.**

surface expulsion. Expulsion occurring at an electrode-to-workpiece contact rather than at the faying surface.

surface preparation. The operations necessary to produce a desired or specified surface condition.

surface roughening. A group of methods for producing irregularities on a surface. See also **dovetailing, groove** and **rotary roughening, rotary roughening, rough threading,** and **threading** and **knurling.**

surfacing. The application by welding, brazing, or thermal spraying of a layer of material to a surface to obtain desired properties or dimensions, as opposed to making a joint. See also **buildup, buttering, cladding,** and **hardfacing.**

surfacing material. The material that is applied to a base metal or substrate during surfacing.

surfacing metal. The metal or alloy that is applied to a base metal or substrate during surfacing.

surfacing weld. A weld applied to a surface, as opposed to making a joint, to obtain desired properties or dimensions. See Figures 15(C) and 30(C).

sweat soldering. A soldering process variation in which workpieces that have been precoated with solder are reheated and assembled into a joint without the use of additional solder.

synchronous timing, *resistance welding.* The initiation of each half cycle of welding transformer primary current on an accurately timed delay with respect to the polarity reversal of the power supply.

T

tab. See **runoff weld tab, starting weld tab,** and **weld tab.**

tacker. A nonstandard term for **tack welder.**

tack weld. A weld made to hold the parts of a weldment in proper alignment until the final welds are made.

taper delay time. The time interval after upslope during which the current is constant. See Figure 52.

taper time. The time interval when current increases or decreases continuously from the welding current to final taper current. See Figure 52.

taps. Connections to a transformer winding that are used to vary the transformer turn ratio, thereby controlling welding voltage and current.

temper time, *resistance welding.* The time following quench time during which a current is passed through the weld for heat treating. See Figure 49.

temporary weld. A weld made to attach a piece or pieces to a weldment for temporary use in handling, shipping, or working on the weldment.

tension test. A test in which a specimen is loaded in tension until failure occurs. See also **reduced section test specimen.**

theoretical electrode force. The force, neglecting friction and inertia, in making spot, seam, or projection welds, available at the electrodes of a resistance welding machine by virtue of the initial force and the theoretical mechanical advantage of the system.

theoretical throat. The distance from the beginning of the joint root perpendicular to the hypotenuse of the largest right triangle that can be inscribed within the cross section of a fillet weld. This dimension is based on the assumption that the root opening is equal to zero. See Figure 25(A), (B), (C), and (D).

thermal cutter. One who performs manual or semiautomatic thermal cutting. Variations of this term are **arc cutter** and **oxygen cutter.**

thermal cutting (TC). A group of cutting processes that severs or removes metal by localized melting, burning, or vaporizing of the workpieces. See also **arc cutting, electron beam cutting, laser beam cutting,** and **oxygen cutting.**

thermal cutting operator. One who operates automatic, mechanized, or robotic thermal cutting equipment. Variations of this term are **arc cutting operator, elec-**

tron beam cutting operator, laser beam cutting operator, and **oxygen cutting operator**.

thermal gouging. A thermal cutting process variation that removes metal by melting or burning the entire removed portion, to form a bevel or groove. See also **arc gouging, backgouging,** and **oxygen gouging**.

thermal spray deposit. The coating or layer of surfacing material applied by a thermal spraying process. See Figure 31(B).

thermal spray deposit density ratio. The ratio of the density of the thermal spray deposit to the theoretical density of the surfacing material, usually expressed as percent of theoretical density.

thermal spray deposit interface. The interface between the thermal spray deposit and the substrate.

thermal spray deposit strength. The tensile strength of a thermal spray deposit.

thermal spray deposit stress. The residual stress in a thermal spray deposit resulting from rapid cooling of molten or semimolten particles as they impinge on the substrate.

thermal sprayer. One who performs semiautomatic thermal spraying. Variations of this term are **arc sprayer, flame sprayer,** and **plasma sprayer**.

| **thermal spraying (THSP).** A group of processes in which finely divided metallic or nonmetallic surfacing materials are deposited in a molten or semimolten condition on a substrate to form a thermal spray deposit. The surfacing material may be in the form of powder, rod, cord, or wire. See also **arc spraying, flame spraying,** and **plasma spraying**.

thermal spraying deposition efficiency. The ratio of the weight of thermal spray deposit to the weight of surfacing material sprayed, expressed in percent.

thermal spraying gun. A device for heating, feeding, and directing the flow of surfacing material.

thermal spraying operator. One who operates automatic, mechanized, or robotic thermal spraying equipment. Variations of this term are **arc spraying operator, flame spraying operator,** and **plasma spraying operator**.

thermal spray pass. A single progression of the thermal spraying gun across the substrate surface.

thermal stress. Stress resulting from nonuniform temperature distribution.

thermite crucible. The vessel in which the thermite reaction takes place.

thermite mixture. A mixture of metal oxide and finely divided aluminum with the addition of alloying metals as required.

thermite mold. A mold formed around the workpieces to receive molten metal.

thermite reaction. The chemical reaction between metal oxide and aluminum that produces superheated molten metal and a slag containing aluminum oxide.

thermite welding (TW). A welding process that produces coalescence of metals by heating them with superheated liquid metal from a chemical reaction between a metal oxide and aluminum, with or without the application of pressure. Filler metal is obtained from the liquid metal.

thermocompression bonding. A nonstandard term for **hot pressure welding**.

threading and knurling, *thermal spraying.* A method of surface roughening in which spiral threads are prepared, followed by upsetting with a knurling tool. See Figure 43(E).

3F. A welding test position designation for a linear fillet weld applied to a joint in which the weld is made in the vertical welding position. See Figure 18(C). ||

3G. A welding test position designation for a linear groove weld applied to a joint in which the weld is made in the vertical welding position. See Figure 17(C). ||

throat area. The area bounded by the physical parts of the secondary circuit in a resistance spot, seam, or projection welding machine. Used to determine the dimensions of a part that can be welded and determine, in part, the secondary impedance of the equipment.

throat crack. See Figure 33.

throat depth. In a resistance spot, seam, or projection welding machine, the distance from the centerline of the electrodes or platens to the nearest point of interference for flat sheets.

throat height. The unobstructed dimension between the arms and throughout the throat depth in a resistance welding machine.

throat length. A nonstandard term when used for **constricting orifice length**. |

throat of a fillet weld. See **actual throat, effective throat,** and **theoretical throat**.

throat of a groove weld. A nonstandard term for **groove weld size**.

throat opening. A nonstandard term for **throat height**.

TIG welding. A nonstandard term for **gas tungsten arc welding.**

tinning. A nonstandard term when used for **precoating.**

tip. See **cutting tip** and **welding tip.**

tip skid. A nonstandard term for **electrode skid.**

T-joint. A joint between two members located approximately at right angles to each other in the form of a T. See Figure 1C, 2C, and 10F.

toe crack. See Figures 32(A) and 33(A).

toe of weld. See **weld toe.**

torch. See **air carbon arc cutting torch, gas tungsten arc cutting torch, gas tungsten arc welding torch, heating torch, oxyfuel gas cutting torch, oxyfuel gas welding torch, plasma arc cutting torch,** and **plasma arc welding torch.**

torch brazing (TB). A brazing process that uses heat from a fuel gas flame.

torch soldering (TS). A soldering process that uses heat from a fuel gas flame.

torch tip. See **welding tip** and **cutting tip.**

transferred arc. A plasma arc established between the electrode of the plasma arc torch and the workpiece.

transverse bend specimen. See **transverse weld test specimen.**

transverse crack. A crack with its major axis oriented approximately perpendicular to the weld axis. See Figure 33(A).

transverse tension specimen. See **transverse weld test specimen.**

transverse weld test specimen. A weld test specimen with its major axis perpendicular to the weld axis. See also **longitudinal weld test specimen.**

travel angle. The angle less than 90 degrees between the electrode axis and a line perpendicular to the weld axis, in a plane determined by the electrode axis and the weld axis. This angle can also be used to partially define the position of guns, torches, rods, and beams. See Figure 21. See also **drag angle, push angle,** and **work angle.**

travel angle, *pipe.* The angle less than 90 degrees between the electrode axis and a line perpendicular to the weld axis at its point of intersection with the extension of the electrode axis, in a plane determined by the electrode axis and a line tangent to the pipe surface at the same point. This angle can also be used to partially define the position of guns, torches, rods, and beams. See Figure 21. See also **drag angle, push angle,** and **work angle.**

travel start delay time. The time interval from arc initiation to the start of the torch, gun, or workpiece travel. See Figure 52.

travel stop delay time. The time interval from beginning of downslope time or crater fill time to shut-off of torch, gun, or workpiece travel. See Figure 52.

tubular joint. A joint between two or more members, at least one of which is tubular.

tungsten electrode. A nonfiller metal electrode used in arc welding, arc cutting, and plasma spraying, made principally of tungsten.

twin carbon arc brazing (TCAB). A brazing process that uses heat from an arc between two carbon electrodes. This is an obsolete or seldom used process.

twin carbon arc welding (CAW-T). A carbon arc welding process variation that uses an arc between two carbon electrodes and no shielding.

2F, *pipe.* A welding test position designation for a circumferential fillet weld applied to a joint in pipe, with its axis approximately vertical, in which the weld is made in the horizontal welding position. See Figure 20(B).

2F, *plate.* A welding test position designation for a linear fillet weld applied to a joint in which the weld is made in the horizontal welding position. See Figure 18(B).

2FR. A welding test position designation for a circumferential fillet weld applied to a joint in pipe, with its axis approximately horizontal, in which the weld is made in the horizontal welding position by rotating the pipe about its axis. See Figure 20(C).

2G, *pipe.* A welding test position designation for a circumferential groove weld applied to a joint in a pipe, with its axis approximately vertical, in which the weld is made in the horizontal welding position. See Figure 19(B).

2G, *plate.* A welding test position designation for a linear groove weld applied to a joint in which the weld is made in the horizontal welding position. See Figure 17(B).

type of joint. See **joint type.**

U

U-groove weld. A type of groove weld. See Figures 8(F) and 9(E).

ultrasonic coupler, *ultrasonic soldering and ultrasonic welding.* Elements through which ultrasonic vibration is transmitted from the transducer to the tip.

ultrasonic soldering (USS). A soldering process variation in which high-frequency vibratory energy is transmitted through molten solder to remove undesirable surface films and thereby promote wetting of the base metal. This operation is usually accomplished without flux.

ultrasonic welding (USW). A solid-state welding process that produces a weld by the local application of high-frequency vibratory energy as the workpieces are held together under pressure.

ultra-speed welding. A nonstandard term for **commutator-controlled welding.**

underbead crack. A crack in the heat-affected zone generally not extending to the surface of the base metal. See Figures 32(B) and 33(A).

undercut. A groove melted into the base metal adjacent to the weld toe or weld root and left unfilled by weld metal. See Figures 32(C) and (D).

underfill. A condition in which the weld face or root surface extends below the adjacent surface of the base metal. See Figures 32(E) and (F).

unmixed zone. A thin boundary layer of weld metal, adjacent to the weld interface, that solidified without mixing with the remaining weld metal. See also **mixed zone.**

uphill, *adv.* Welding with an upward progression.

upset. Bulk deformation resulting from the application of pressure in welding. The upset may be measured as a percent increase in interface area, a reduction in length, a percent reduction in lap joint thickness, or a reduction in cross wire weld stack height.

upset butt welding. A nonstandard term for **upset welding.**

upset distance. The total reduction in the axial length of the workpieces from the initial contact to the completion of the weld. In flash welding the upset distance is equal to the platen movement from the end of flash time to the end of upset.

upset force. The force exerted at the faying surfaces during upsetting.

upset time. The time during upsetting.

upset welding (UW). A resistance welding process that produces coalescence over the entire area of faying surfaces or progressively along a butt joint by the heat obtained from the resistance to the flow of welding current through the area where those surfaces are in contact. Pressure is used to complete the weld. See Figures 15(A), 31(C), and 51. See also **high-frequency upset welding** and **induction upset welding.**

upslope time. See **automatic arc welding upslope time** and **resistance welding upslope time.**

V

vacuum brazing. A nonstandard term for various brazing processes that take place in a chamber or retort below atmospheric pressure.

vacuum plasma spraying (VPSP). A thermal spraying process variation using a plasma spraying gun confined to a stable enclosure that is partially evacuated.

vertical-down. A nonstandard term for **downhill.**

vertical position. See **vertical welding position.**

vertical position, *pipe welding.* A nonstandard term for 2G in pipe.

vertical welding position. The welding position in which the weld axis, at the point of welding, is approximately vertical, and the weld face lies in an approximately vertical plane. See Figures 16A, 16B, 16C, 17(C), and 18(C).

vertical-up. A nonstandard term for **uphill.**

V-groove weld. A type of groove weld. See Figures 8(C), 8(D) and 9(C).

voltage regulator. An automatic electrical control device for maintaining a constant voltage supply to the primary of a welding transformer.

W

water wash. The forcing of exhaust air and fumes from a spray booth through water so that the vented air is free of thermal sprayed particles or fumes.

wave soldering (WS). An automatic soldering process where workpieces are passed through a wave of molten solder. See also **dip soldering.**

wax pattern, *thermite welding.* Wax molded around the workpieces to the form desired for the completed weld.

weave bead. A type of weld bead made with transverse oscillation. See Figure 22(B).

weld. A localized coalescence of metals or nonmetals produced either by heating the materials to the welding temperature, with or without the application of pressure, or by the application of pressure alone and with or without the use of filler material.

weldability. The capacity of material to be welded under the imposed fabrication conditions into a specific, suitably designed structure and to perform satisfactorily in the intended service.

weld axis. A line through the length of the weld, perpendicular to and at the geometric center of its cross section. See Figures 16A, 16B, and 21.

weld bead. A weld resulting from a pass. See Figures 22, 23(D), and 23(E). See also **stringer bead** and **weave bead.**

weld bonding. A resistance spot welding process variation in which the spot weld strength is augmented by adhesive at the faying surfaces.

weld brazing. A joining method that combines resistance welding with brazing.

weld crack. A crack located in the weld metal or heat-affected zone. See Figure 33.

welder. One who performs manual or semiautomatic welding.

welder certification. Written verification that a welder has produced welds meeting a prescribed standard of welder performance.

welder performance qualification. The demonstration of a welder's ability to produce welds meeting prescribed standards.

welder registration. The act of registering a welder certification or a photostatic copy of the welder certification.

weld face. The exposed surface of a weld on the side from which welding was done. See Figures 24(A) and (E).

weld face underfill. See **underfill.** See Figures 32(E) and (F).

weld gage. A device designed for measuring the shape and size of welds.

weld groove. A channel in the surface of a workpiece or an opening between two joint members that provides space to contain a weld. See Figures 8, 9 and 13(B).

welding. A joining process that produces coalescence of materials by heating them to the welding temperature, with or without the application of pressure or by the application of pressure alone, and with or without the use of filler metal. See also the Master Chart of Welding and Allied Processes.

welding arc. A controlled electrical discharge between the electrode and the workpiece that is formed and sustained by the establishment of a gaseous conductive medium, called an *arc plasma.*

welding blowpipe. A nonstandard term for **oxyfuel gas welding torch.**

welding current. See **automatic arc welding current** and **resistance welding current.**

welding cycle. The complete series of events involved in the making of a weld. See Figures 49, 50, 52, and 53.

welding electrode. A component of the welding circuit through which current is conducted and that terminates at the arc, molten conductive slag, or base metal. See also **arc welding electrode, bare electrode, carbon electrode, composite electrode, covered electrode, electroslag welding electrode, emissive electrode, flux cored electrode, lightly coated electrode, metal cored electrode, metal electrode, resistance welding electrode, stranded electrode,** and **tungsten electrode.**

welding filler metal. The metal or alloy to be added in making a weld joint that alloys with the base metal to form weld metal in a fusion welded joint.

welding force. See **dynamic electrode force, electrode force, forge force, friction welding force, static electrode force, theoretical electrode force,** and **upset force.**

welding generator. A generator used for supplying current for welding.

welding ground. A nonstandard and incorrect term for **workpiece connection.**

welding head. The part of a welding machine in which a welding gun or torch is incorporated.

welding helmet. A device equipped with a filter plate designed to be worn on the head to protect eyes, face, and neck from arc radiation, radiated heat, spatter or other harmful matter expelled during some welding and cutting processes.

welding hood. A nonstandard term for **welding helmet.**

welding leads. The workpiece lead and electrode lead of an arc welding circuit. See Figure 34.

welding machine. Equipment used to perform the welding operation. For example, spot welding machine, arc welding machine, and seam welding machine.

welding operator. One who operates adaptive control, automatic, mechanized, or robotic welding equipment.

welding position. The relationship between the weld pool, joint, joint members, and welding heat source during welding. See also **flat welding position, horizontal welding position, overhead welding position,** and **vertical welding position**. See Figures 16–20.

welding power source. An apparatus for supplying current and voltage suitable for welding. See also **constant current power source, constant voltage power source, welding generator, welding rectifier,** and **welding transformer**.

welding procedure. The detailed methods and practices involved in the production of a weldment. See also **welding procedure specification**.

welding procedure qualification record (WPQR). A record of welding variables used to produce an acceptable test weldment and the results of tests conducted on the weldment to qualify a welding procedure specification.

welding procedure specification (WPS). A document providing the required welding variables for a specific application to assure repeatability by properly trained welders and welding operators.

welding rectifier. A device in a welding power source for converting alternating current to direct current.

welding rod. A form of welding filler metal, normally packaged in straight lengths, that does not conduct the welding current. See Figure 36.

welding schedule. A written statement, usually in tabular form, specifying values of parameters and the welding sequence for performing a welding operation.

welding sequence. The order of making welds in a weldment.

welding symbol. A graphical representation of a weld.

welding technique. The details of a welding procedure that are controlled by the welder or welding operator.

welding test position designation. A symbol representation for a fillet weld or a groove weld, the joint orientation and the welding test position. See **1F, 2F, 2FR, 3F, 4F, 5F, 6F, 1G, 2G, 3G, 4G, 5G, 6G**, and **6GR**.

welding tip. That part of an oxyfuel gas welding torch from which the gases issue.

welding torch. See **gas tungsten arc welding torch, oxyfuel gas welding torch**, and **plasma arc welding torch**.

welding transformer. A transformer used for supplying current for welding.

welding voltage. See **arc voltage**.

welding wheel. A nonstandard term for **resistance welding electrode**.

welding wire. A form of welding filler metal, normally packaged as coils or spools, that may or may not conduct electrical current depending upon the welding process with which it is used. See also **welding electrode** and **welding rod**.

weld interface. The interface between weld metal and base metal in a fusion weld, between base metals in a solid-state weld without filler metal, or between filler metal and base metal in a solid-state weld with filler metal. See Figures 30 and 31.

weld interval, *resistance welding*. The total of all heat and cool times, and upslope time, used in making one multiple-impulse weld. See Figure 49. See also **weld time**.

weld joint mismatch. Misalignment of the joint members. See Figure 13(C).

weld line. A nonstandard term for **weld interface**.

weldment. An assembly whose component parts are joined by welding.

weld metal. The portion of a fusion weld that has been completely melted during welding. See Figure 24(G). See also **mixed zone** and **unmixed zone**.

weld metal area. The area of weld metal as measured on the cross section of a weld. See Figure 24(G).

weld metal crack. See Figure 33.

weldor. A nonstandard term for **welder**.

weld pass. A single progression of welding along a joint. The result of a pass is a weld bead or layer.

weld pass sequence. The order in which the weld passes are made. See **longitudinal sequence** and **cross-sectional sequence**.

weld penetration. A nonstandard term for **joint penetration** and **root penetration**.

weld pool. The localized volume of molten metal in a weld prior to its solidification as weld metal.

weld puddle. A nonstandard term for **weld pool**.

weld recognition. A function of an adaptive control that determines changes in the shape of the weld pool or the weld metal during welding, and directs the welding machine to take appropriate action. See also **joint recognition** and **joint tracking**.

weld reinforcement. Weld metal in excess of the quantity required to fill a joint. See also **face reinforcement** and **root reinforcement**.

weld root. The points, shown in a cross section, at which the root surface intersects the base metal surfaces. See Figure 24.

weld seam. A nonstandard term for **joint, seam weld, weld,** or **weld joint.**

weld size. See **edge weld size, fillet weld size, groove weld size, plug weld size, projection weld size, seam weld size, slot weld size** and **spot weld size.**

weld symbol. A graphical character connected to the welding symbol indicating the type of weld.

weld tab. Additional material that extends beyond either end of the joint, on which the weld is started or terminated. See **runoff weld tab** and **starting weld tab.**

weld throat. See **actual throat, effective throat,** and **theoretical throat.**

weld time. See **automatic arc welding weld time** and **resistance welding weld time.**

weld toe. The junction of the weld face and the base metal. See Figures 24(A) and (E).

weld voltage. See **arc voltage.**

wetting. The phenomenon whereby a liquid filler metal or flux spreads and adheres in a thin continuous layer on a solid base metal.

whipping. A manual welding technique in which the arc or flame is manipulated to alternate backwards and forwards as it progresses along the weld path.

wiped joint. A joint made with solder having a wide melting range and with the heat supplied by the molten solder poured onto the joint. The solder is manipulated with a hand-held cloth or paddle so as to obtain the required size and contour.

wire feed speed. The rate at which wire is consumed in arc cutting, thermal spraying, or welding.

wire flame spraying (FLSP-W). A thermal spraying process variation in which the surfacing material is in wire form.

wire straightener. A device used for controlling the cast and helix of coiled wire to enable it to be easily fed through the wire feed system.

work angle. The angle less than 90 degrees between a line perpendicular to the major workpiece surface and a plane determined by the electrode axis and the weld axis. In a T-joint or a corner joint, the line is perpendicular to the nonbutting member. This angle can also be used to partially define the position of guns, torches, rods, and beams. See Figure 21. See also **drag angle, push angle,** and **travel angle.**

work angle, *pipe.* The angle less than 90° between a line that is perpendicular to the cylindrical pipe surface at the point of intersection of the weld axis and the extension of the electrode axis, and a plane determined by the electrode axis and a line tangent to the pipe at the same point. In a T-joint, the line is perpendicular to the nonbutting member. This angle can also be used to partially define the position of guns, torches, rods, and beams. See Figure 21. See also **drag angle, push angle,** and **travel angle.**

work coil. See **induction work coil.**

work connection. A nonstandard term for **workpiece connection.**

work lead. A nonstandard term for **workpiece lead.**

workpiece. The part that is welded, brazed, soldered, thermal cut, or thermal sprayed.

workpiece connection. The connection of the workpiece lead to the workpiece. See Figure 34.

workpiece lead. The electrical conductor between the arc welding current source and workpiece connection. See Figure 34.

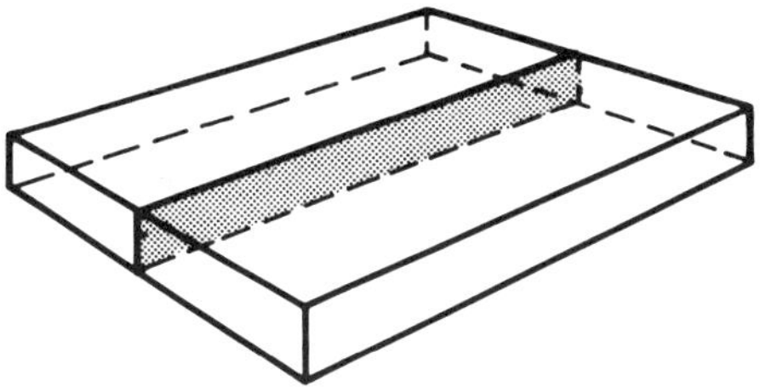

APPLICABLE WELDS

BEVEL-GROOVE	U-GROOVE
FLARE-BEVEL-GROOVE	V-GROOVE
FLARE-V-GROOVE	EDGE-FLANGE
J-GROOVE	BRAZE
SQUARE-GROOVE	

(A) BUTT JOINT

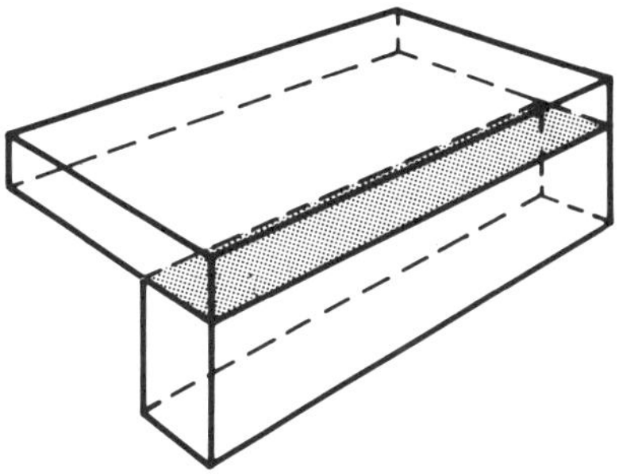

APPLICABLE WELDS

FILLET	V-GROOVE
BEVEL-GROOVE	PLUG
FLARE-BEVEL-GROOVE	SLOT
FLARE-V-GROOVE	SPOT
J-GROOVE	SEAM
SQUARE-GROOVE	PROJECTION
U-GROOVE	BRAZE

(B) CORNER JOINT

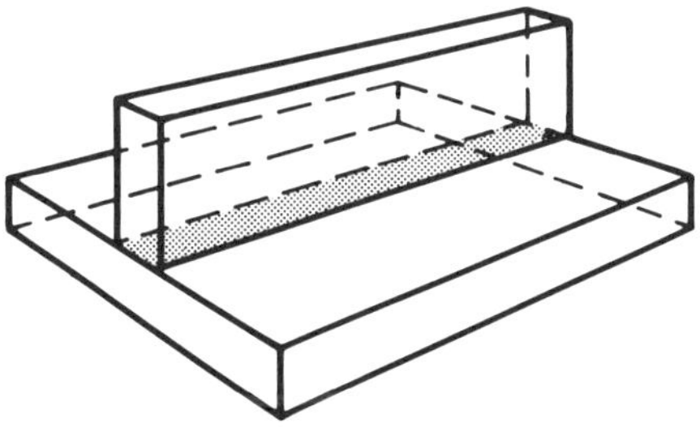

APPLICABLE WELDS

FILLET	SLOT
BEVEL-GROOVE	SPOT
FLARE-BEVEL-GROOVE	SEAM
J-GROOVE	PROJECTION
SQUARE-GROOVE	BRAZE
PLUG	

(C) T-JOINT

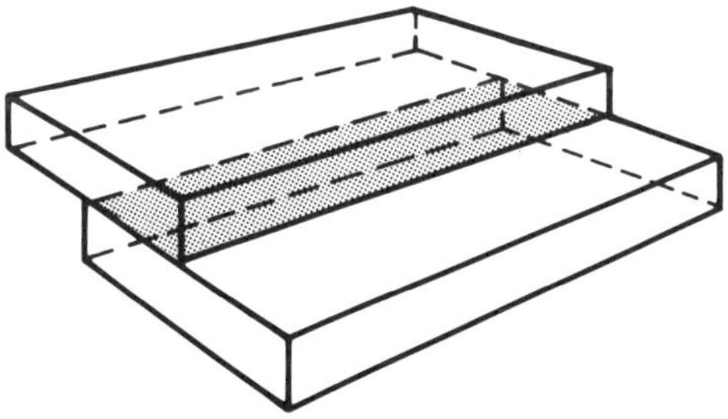

APPLICABLE WELDS

FILLET	SLOT
BEVEL-GROOVE	SPOT
FLARE-BEVEL-GROOVE	SEAM
J-GROOVE	PROJECTION
PLUG	BRAZE

(D) LAP JOINT

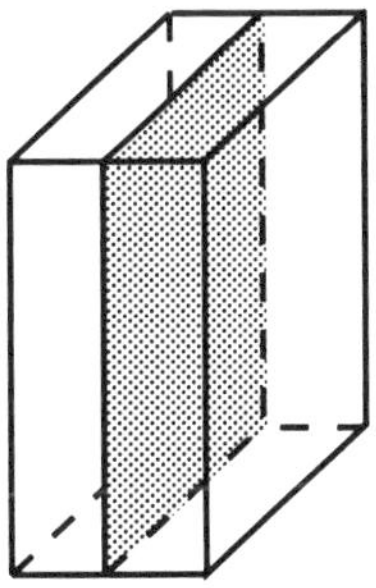

APPLICABLE WELDS

BEVEL-GROOVE	U-GROOVE
FLARE-BEVEL-GROOVE	V-GROOVE
FLARE-V-GROOVE	EDGE
J-GROOVE	SEAM
SQUARE-GROOVE	

(E) EDGE JOINT

Figure 1 — Joint Types

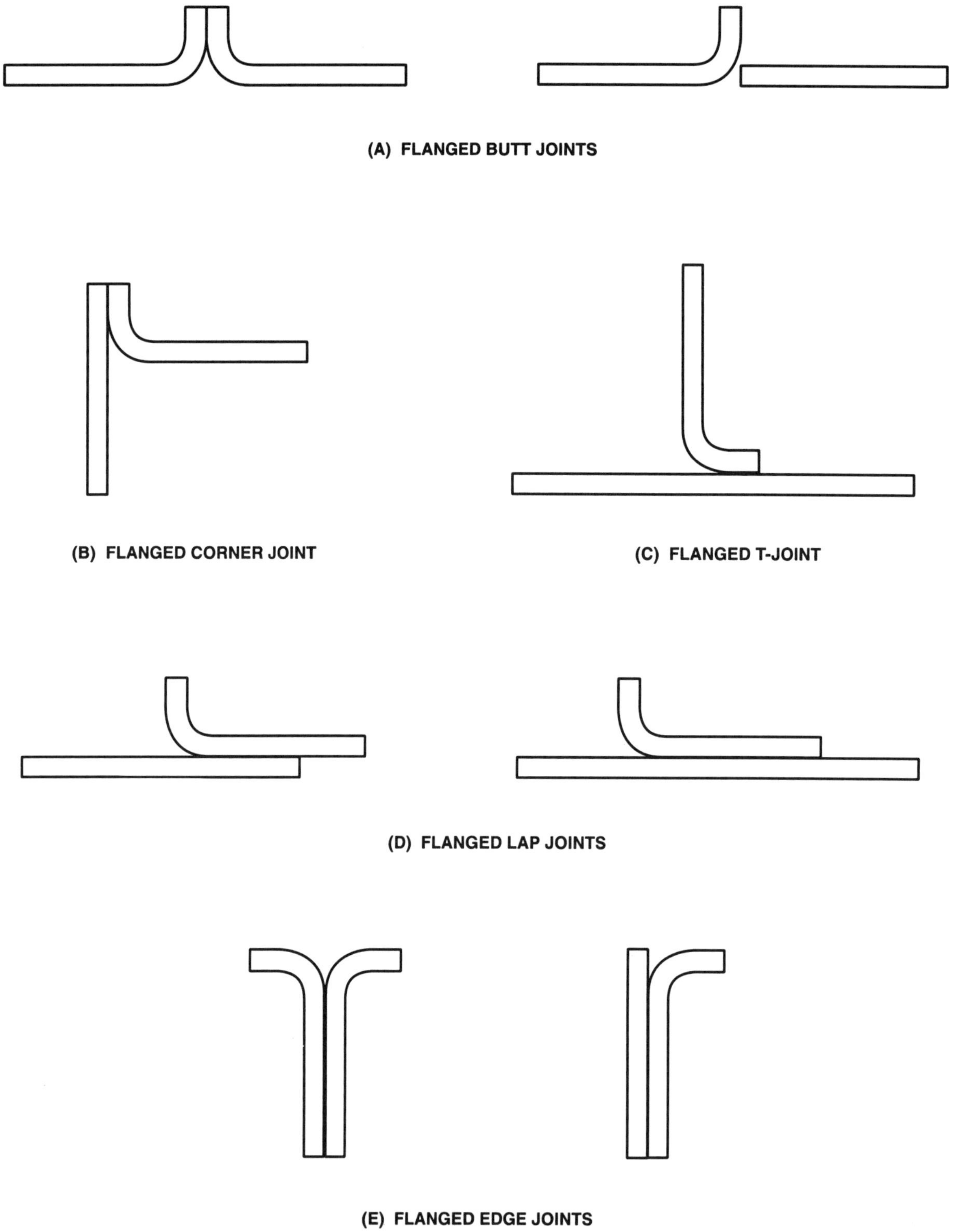

Figure 2 — Flanged Joints

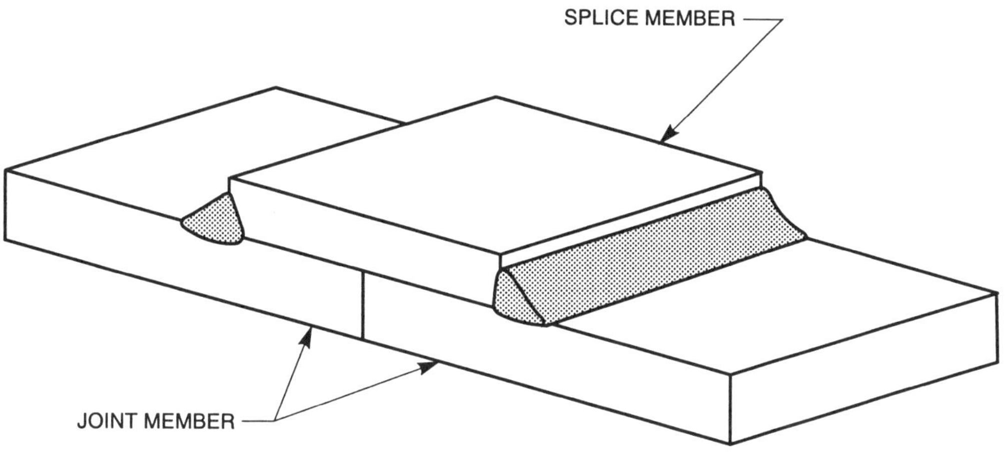

(A) SINGLE-SPLICED BUTT JOINT

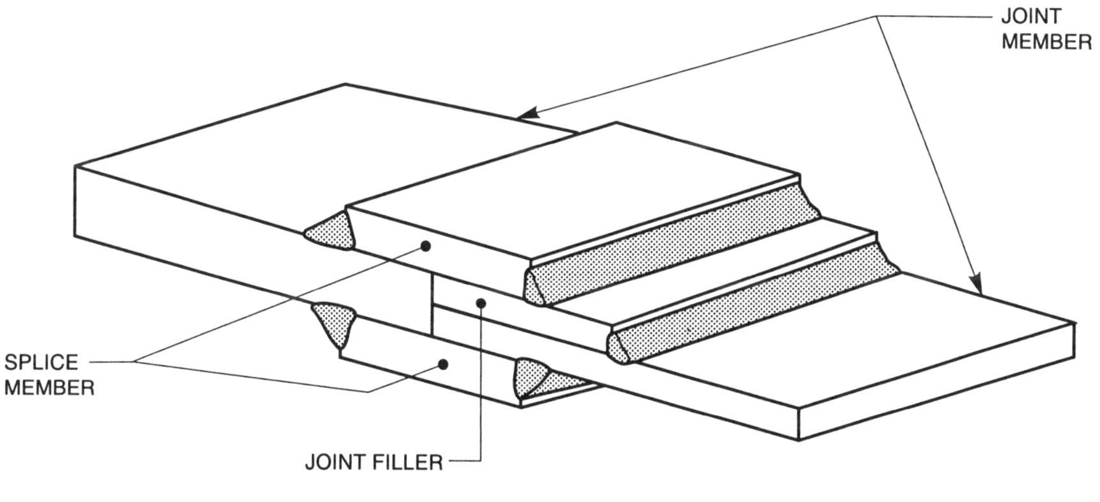

(B) DOUBLE-SPLICED EDGE JOINT WITH JOINT FILLER

Figure 3 — Spliced Butt Joints

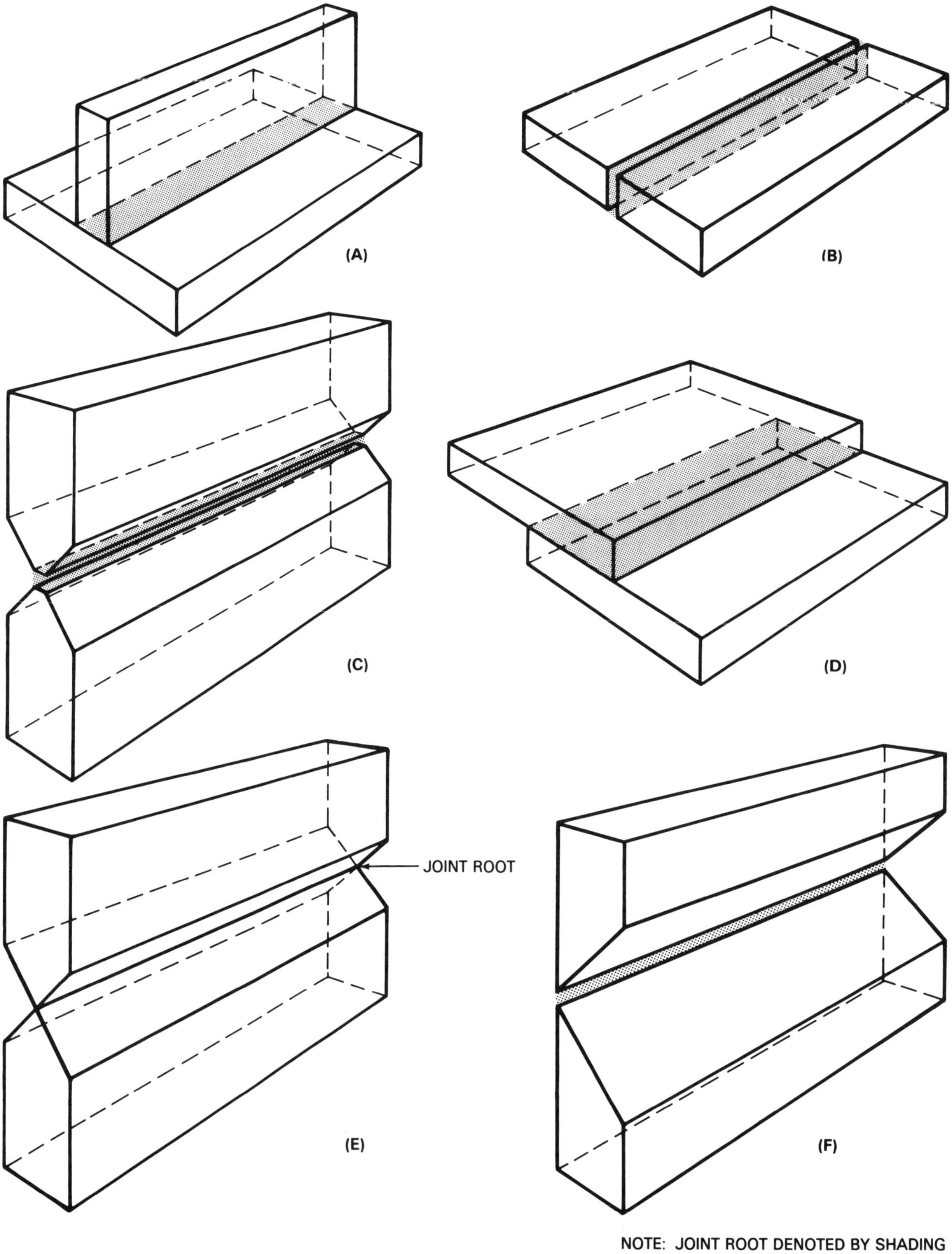

Figure 4 — Joint Root

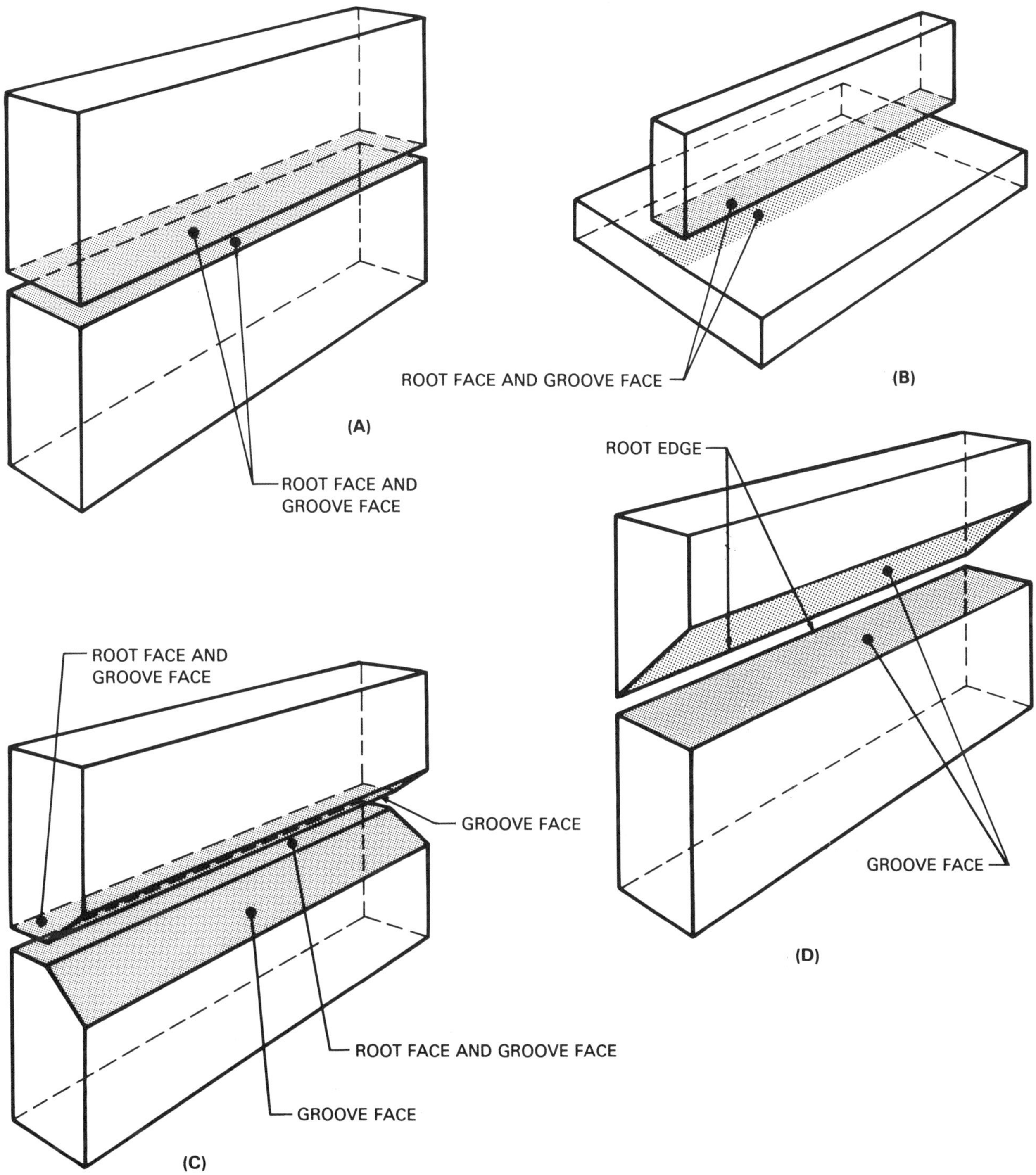

Figure 5 — Groove Face, Root Edge, and Root Face

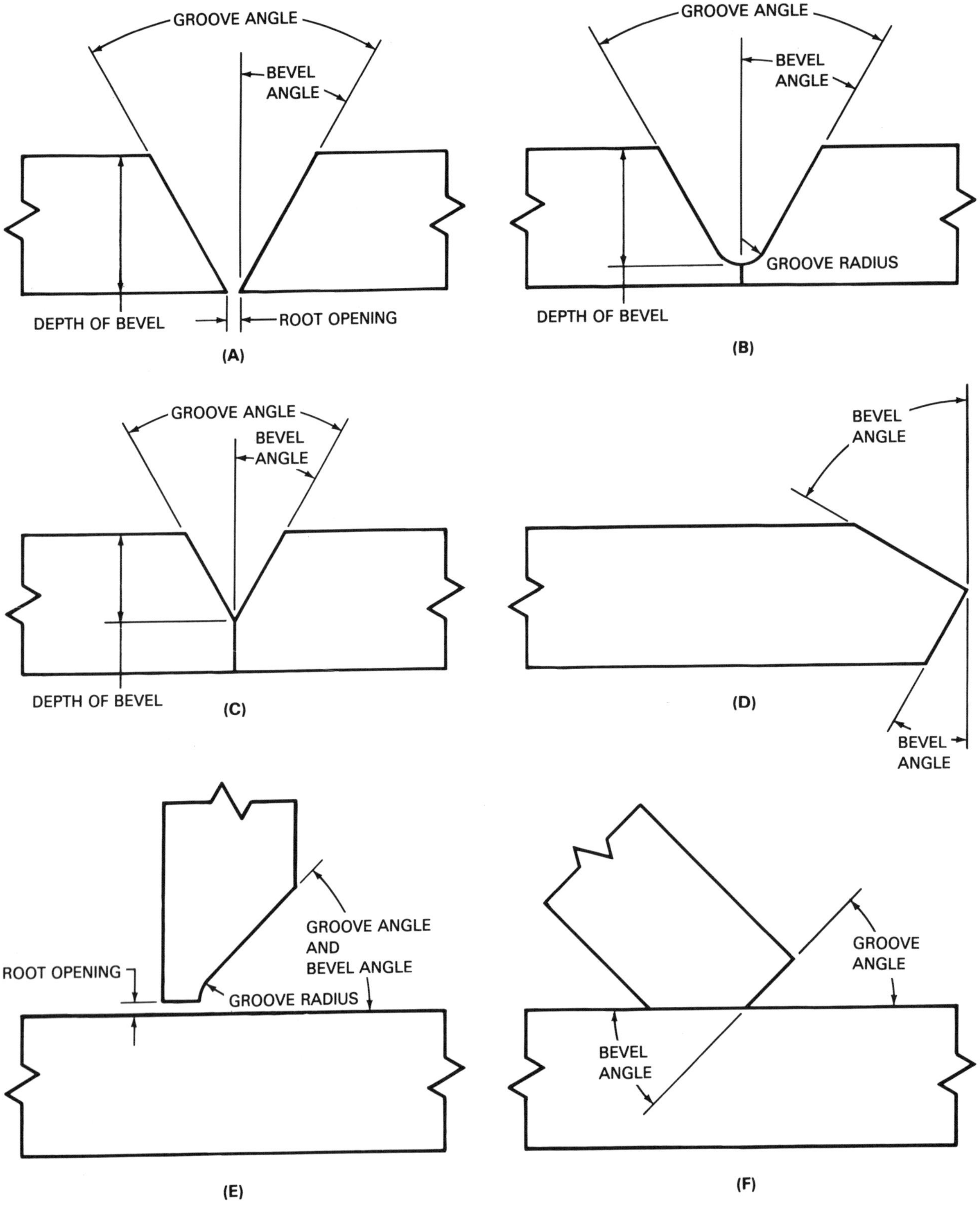

Figure 6 — Bevel Angle, Depth of Bevel, Groove Angle, Groove Radius, and Root Opening

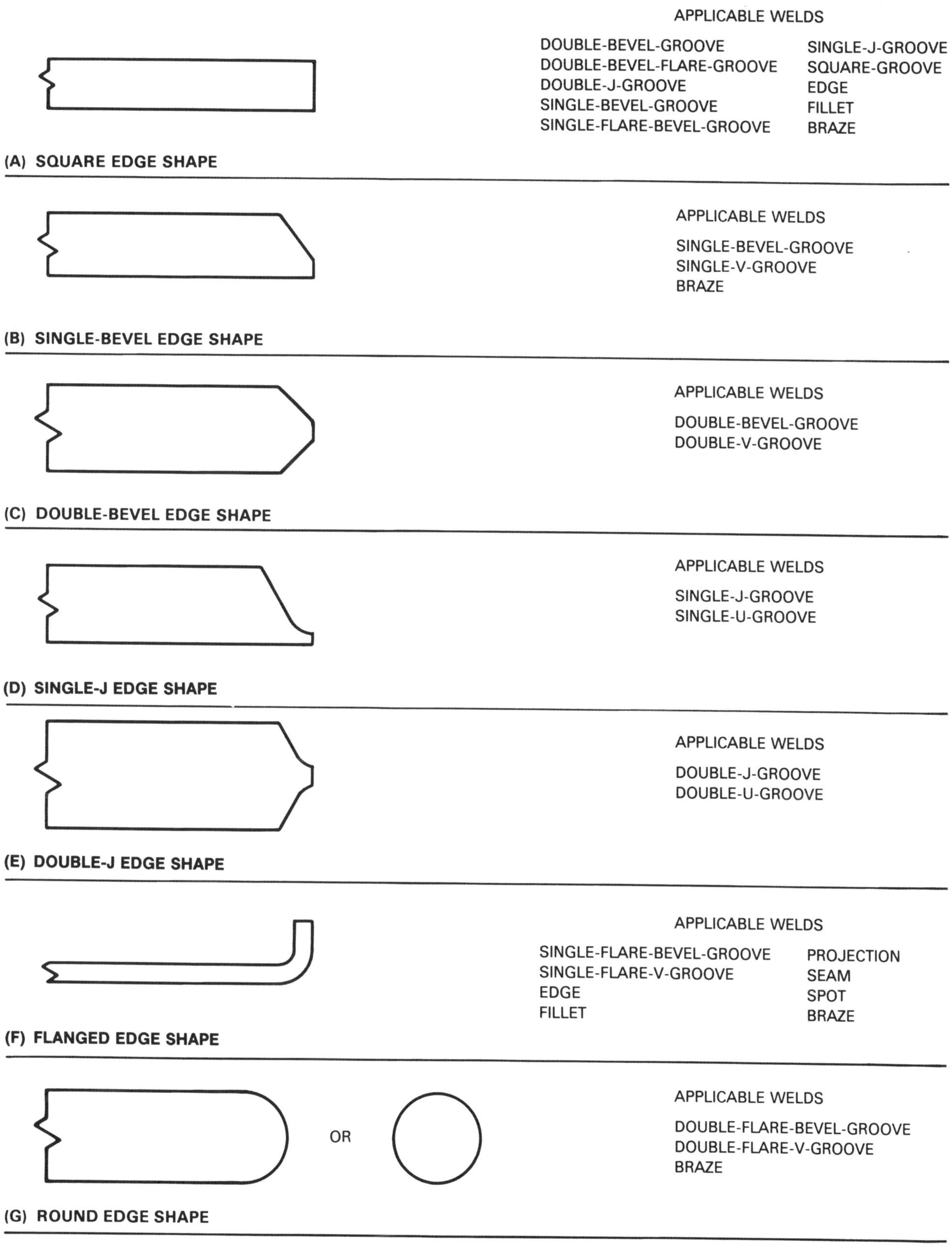

Figure 7 — Edge Shapes

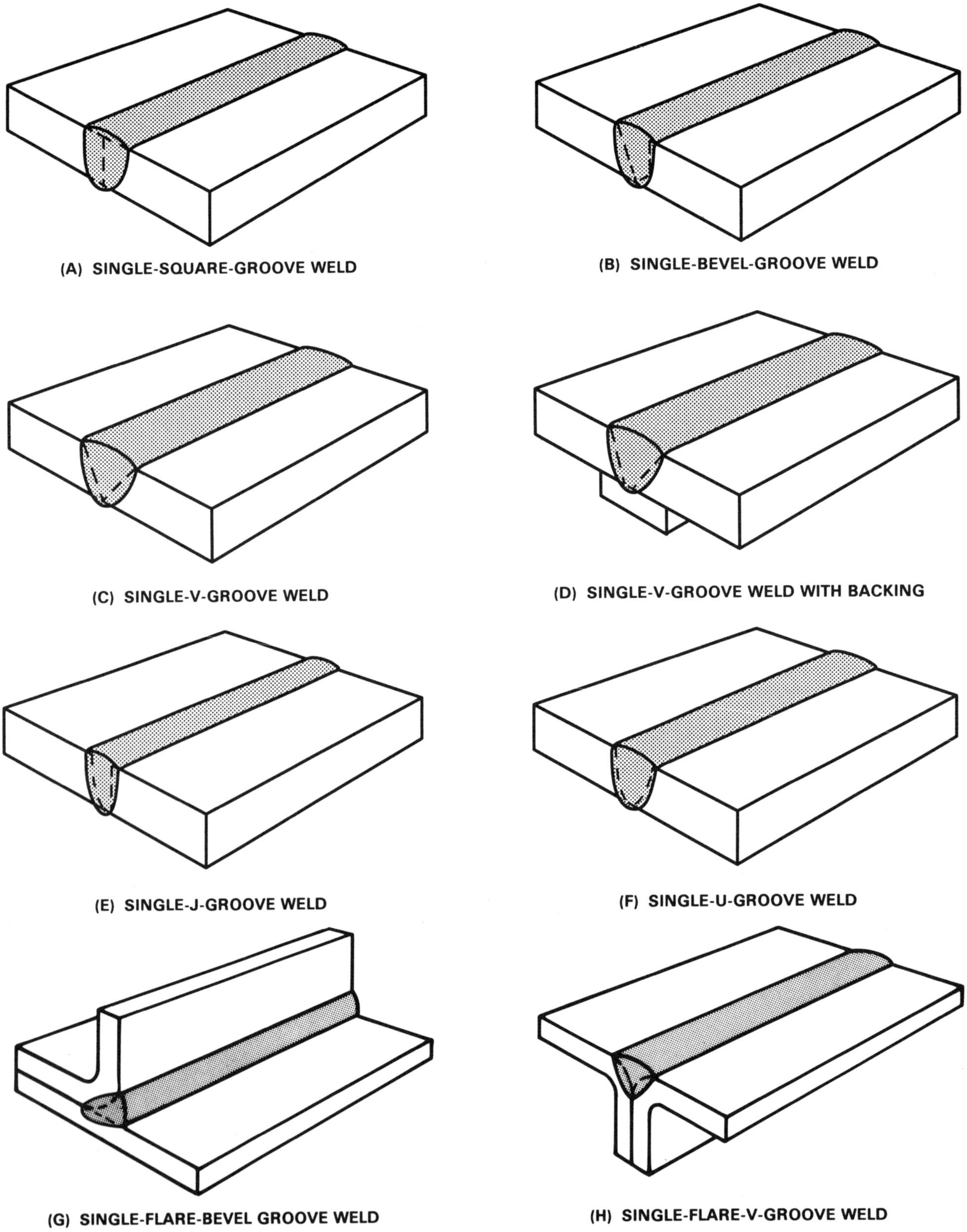

Figure 8 — Single-Groove Weld Joints

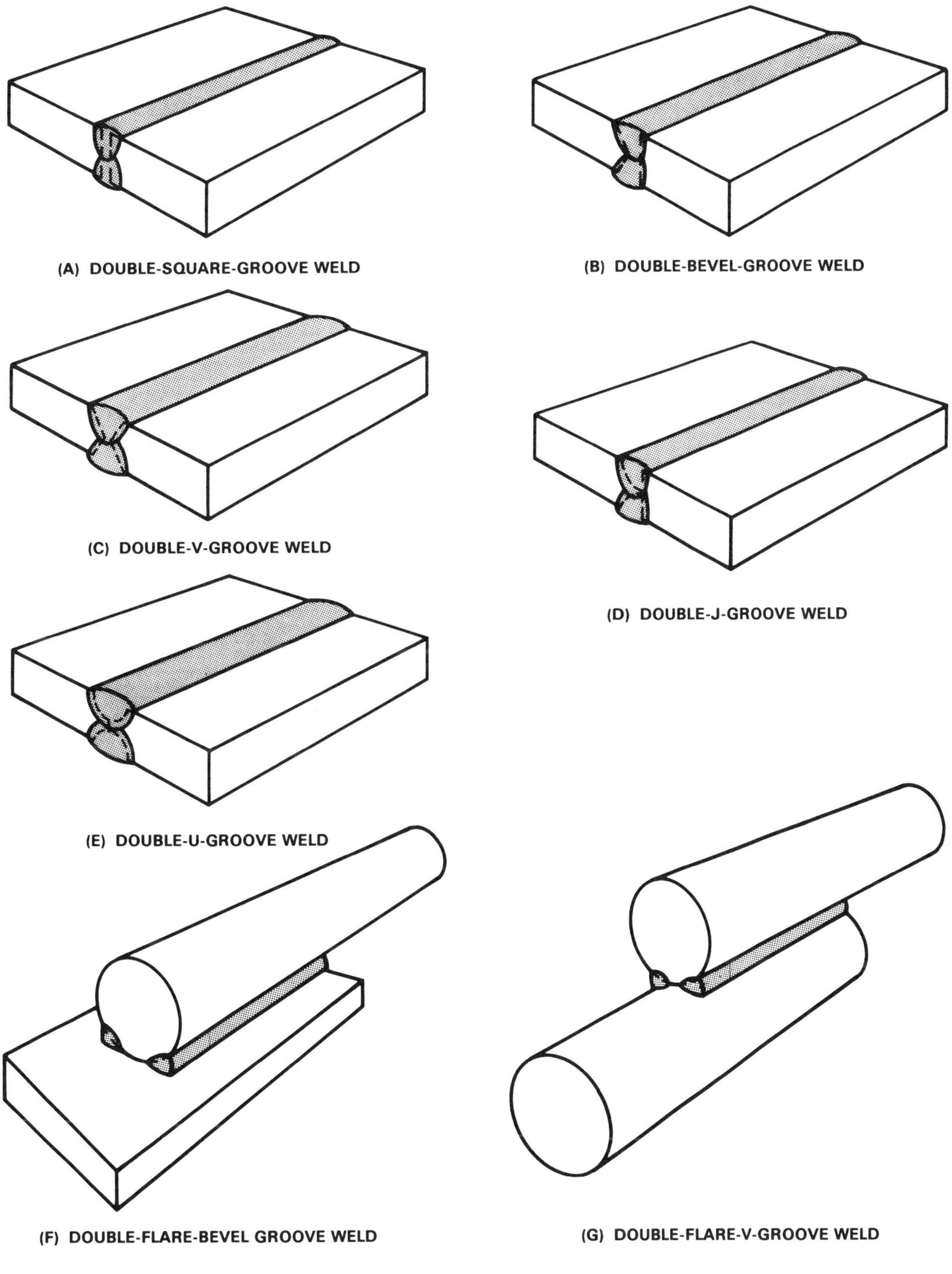

Figure 9 — Double-Groove Weld Joints

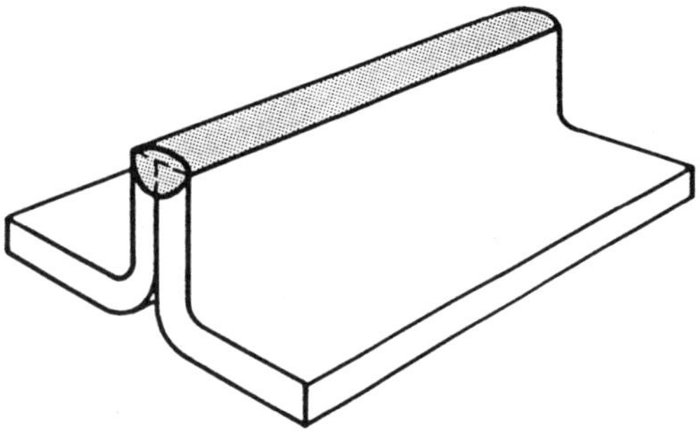

(A) EDGE WELD IN A FLANGED BUTT JOINT

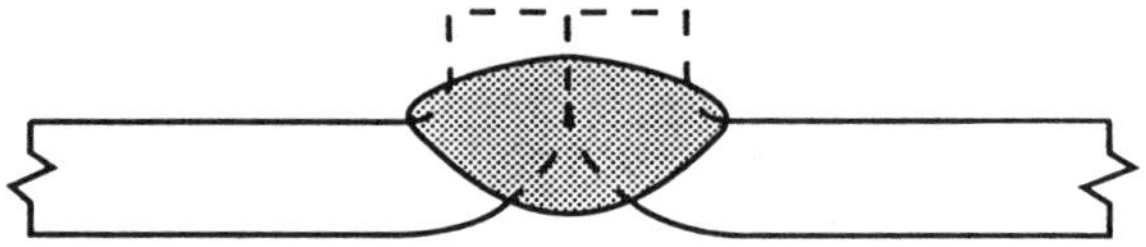

(B) EDGE WELD WITH MELT-THROUGH IN A FLANGED BUTT JOINT

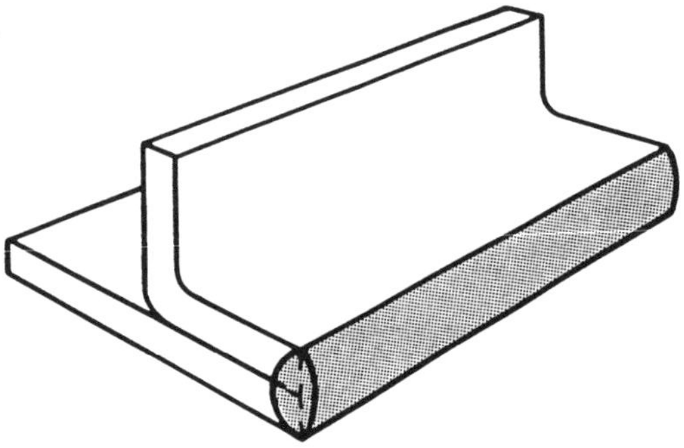

(C) EDGE WELD IN A FLANGED CORNER JOINT

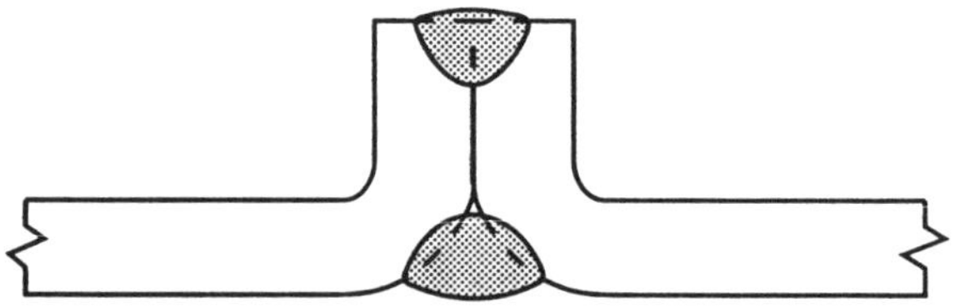

(D) SQUARE-GROOVE WELD AND FLARE-V-GROOVE WELD IN A FLANGED BUTT JOINT

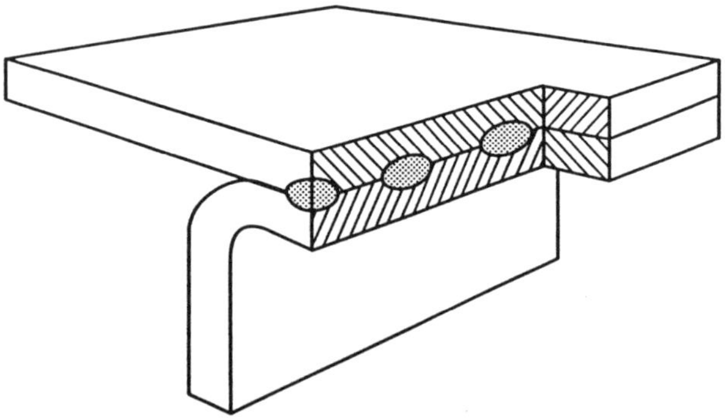

(E) RESISTANCE SPOT WELDS IN A FLANGED CORNER JOINT

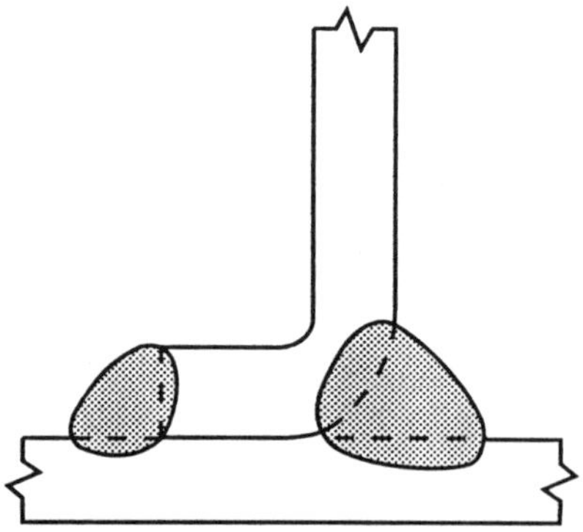

(F) FILLET WELD AND FLARE-BEVEL-GROOVE WELD IN A FLANGED T-JOINT

Figure 10 — Welds in Flanged Joints

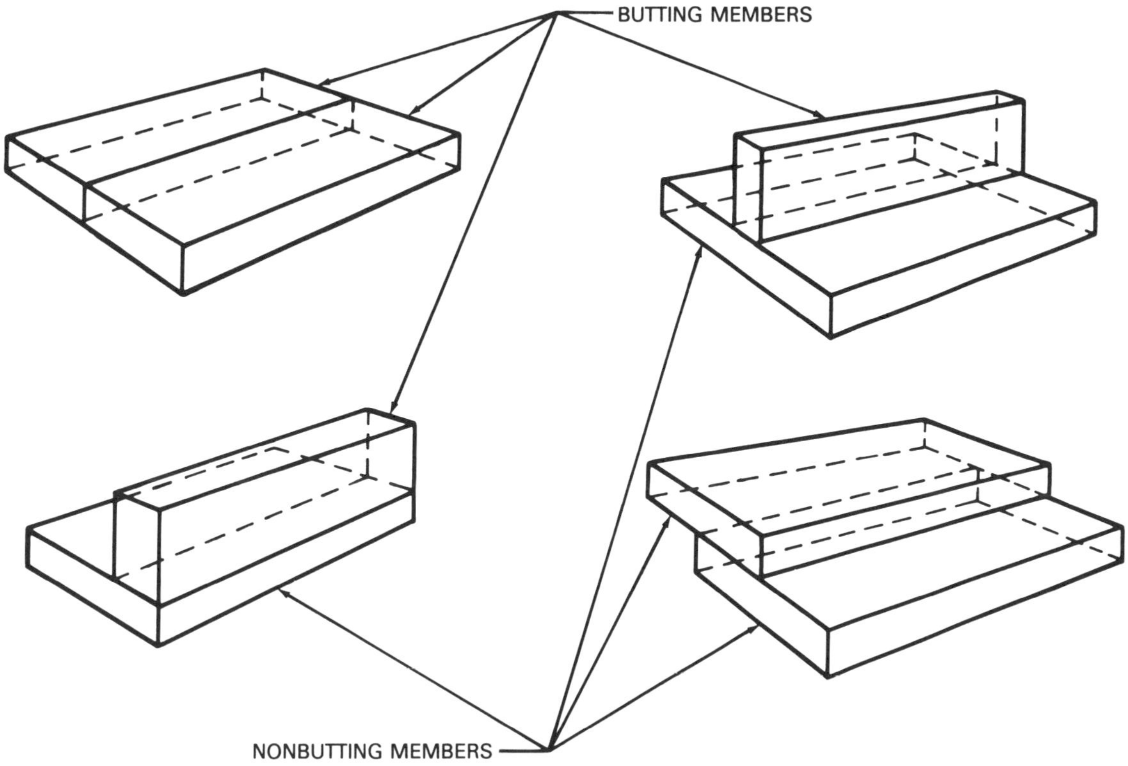

Figure 11 — Butting and Nonbutting Members

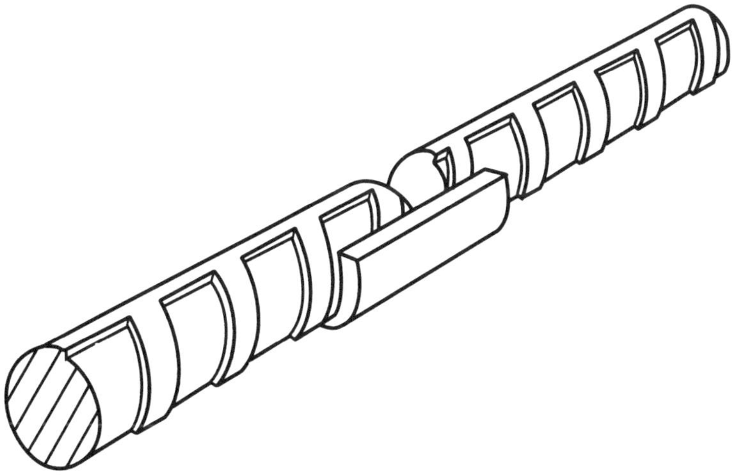

Figure 12 — Split Pipe Backing

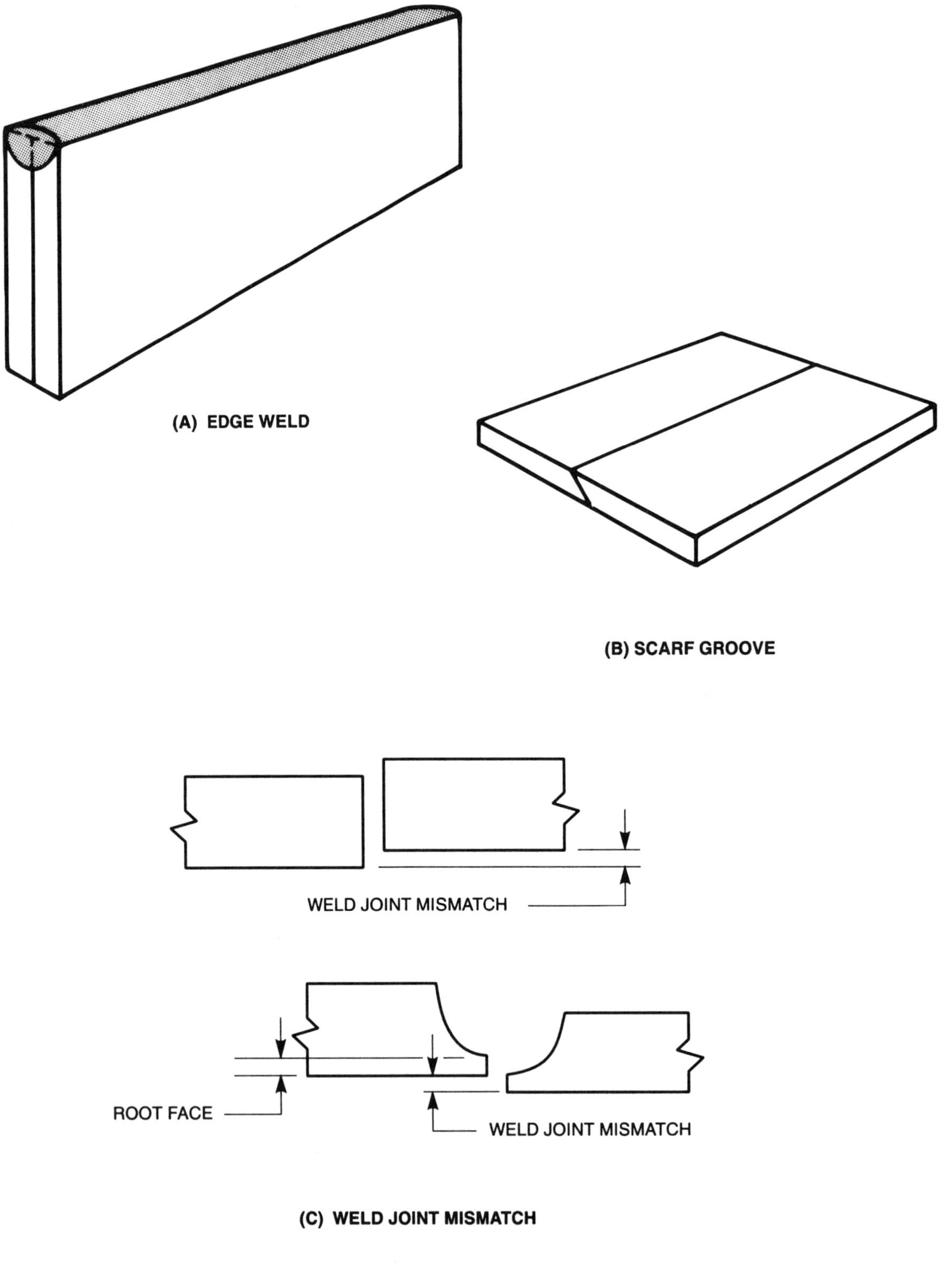

Figure 13 — Edge Weld, Scarf Groove, and Weld Joint Mismatch

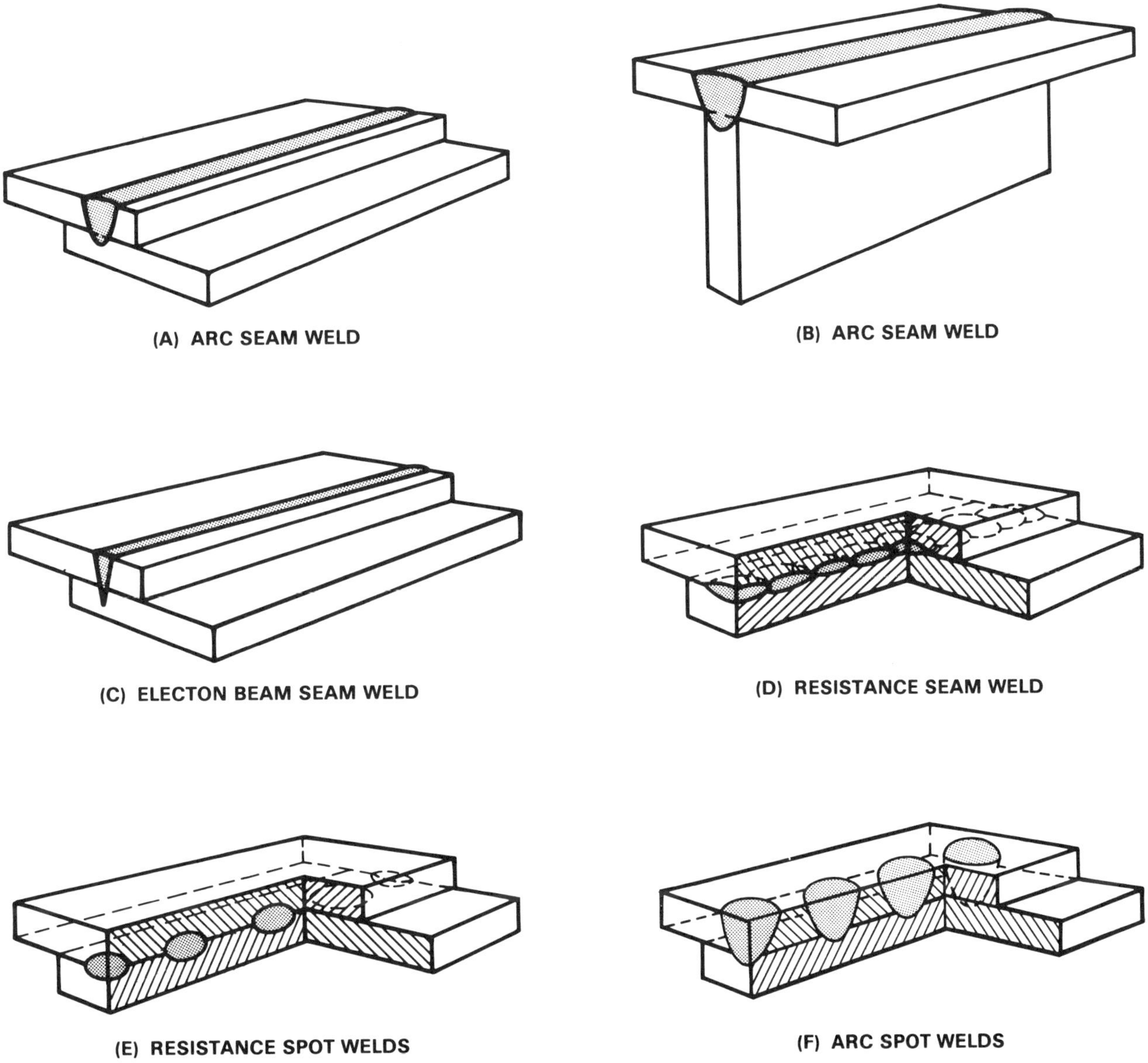

Figure 14 — Seam Welds and Spot Welds

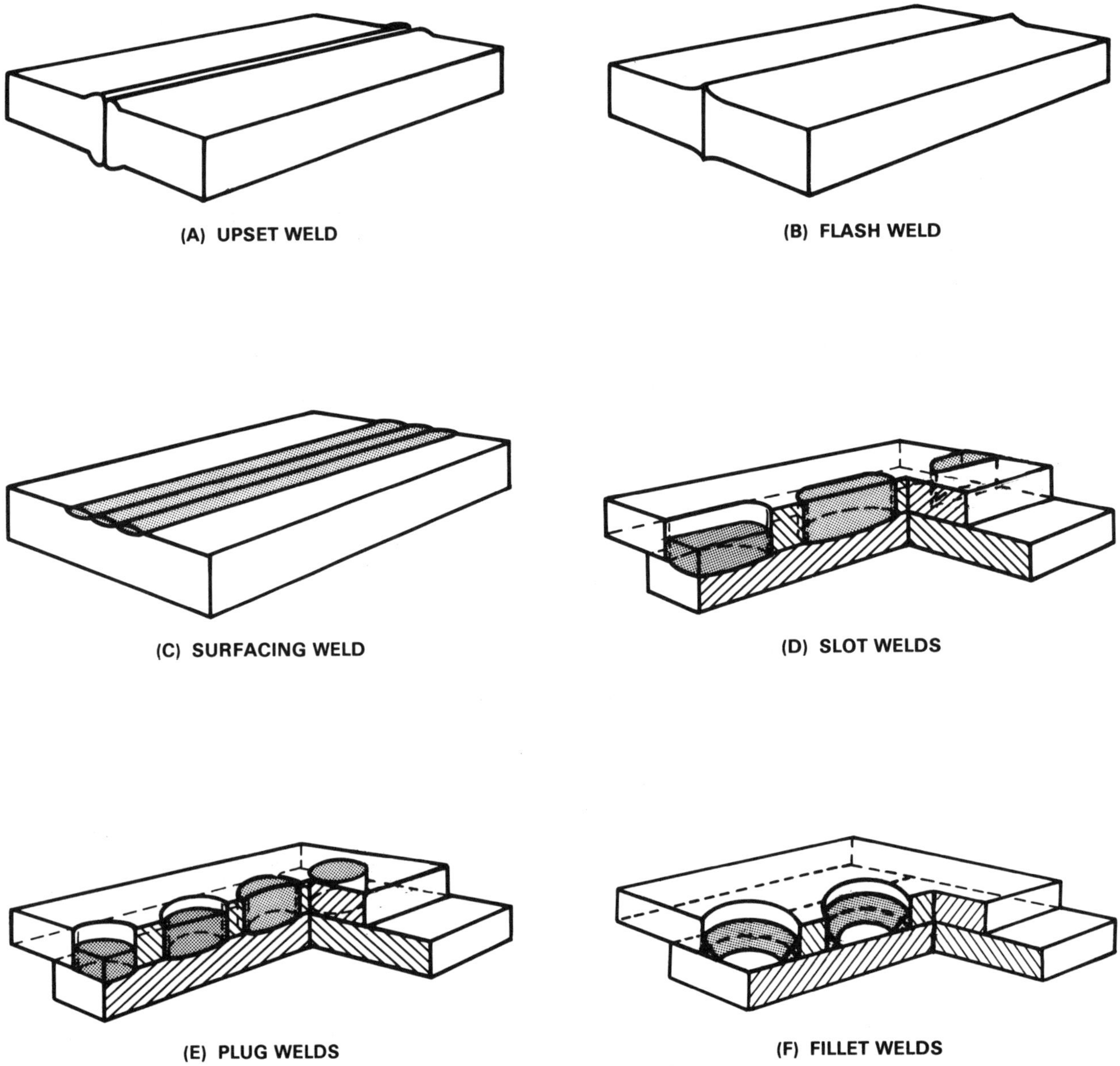

Figure 15 — Various Weld Types

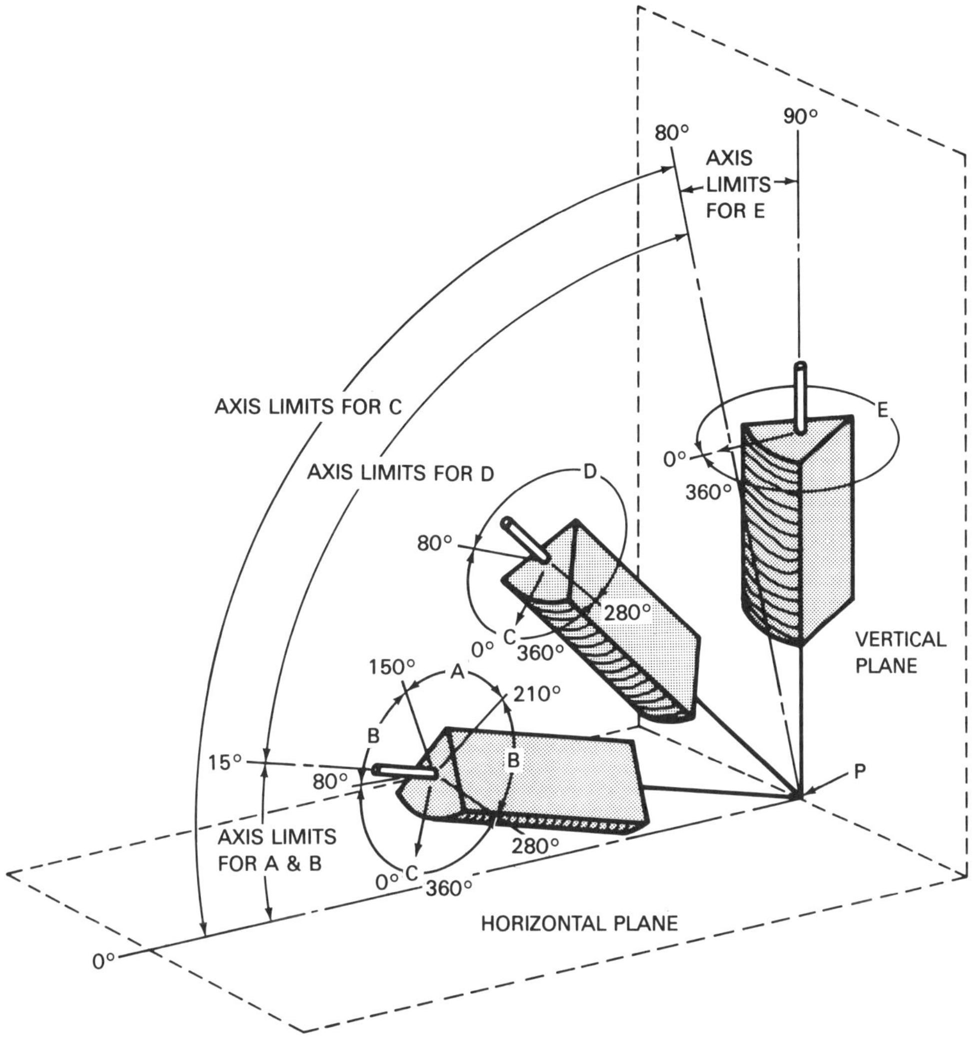

TABULATION OF POSITIONS OF GROOVE WELDS			
POSITION	DIAGRAM REFERENCE	INCLINATION OF AXIS	ROTATION OF FACE
FLAT	A	0° to 15°	150° to 210°
HORIZONTAL	B	0° to 15°	80° to 150° 210° to 280°
OVERHEAD	C	0° to 80°	0° to 80° 280° to 360°
VERTICAL	D E	15° to 80° 80° to 90°	80° to 280° 0° to 360°

Notes:
1. The horizontal reference plane is always taken to lie below the weld under consideration.
2. The inclination of the weld axis is measured from the horizontal reference plane toward the vertical reference plane.
3. The angle of rotation of the weld face is determined by a line perpendicular to the weld face at its center which passes through the weld axis. The reference position (0°) of rotation of the weld face invariably points in the direction opposite to that in which the weld axis angle increases. When looking at point P, the angle of rotation of the weld face is measured in a clockwise direction from the reference position (0°).

Figure 16A — Welding Position Diagram for Groove Welds in Plate

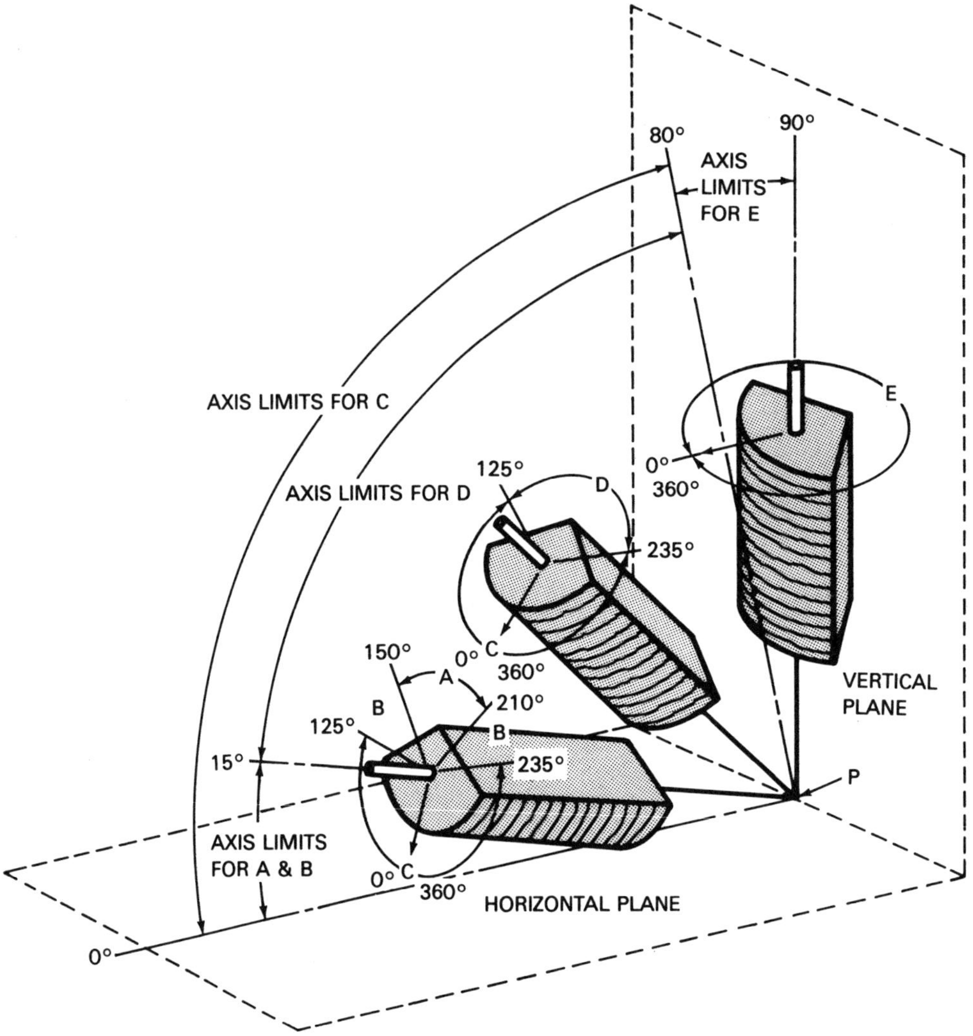

TABULATIONS OF POSITIONS OF FILLET WELDS			
POSITION	DIAGRAM REFERENCE	INCLINATION OF AXIS	ROTATION OF FACE
FLAT	A	0° to 15°	150° to 210°
HORIZONTAL	B	0° to 15°	125° to 150° 210° to 235°
OVERHEAD	C	0° to 80°	0° to 125° 235° to 360°
VERTICAL	D E	15° to 80° 80° to 90°	125° to 235° 0° to 360°

Notes:
1. The horizontal reference plane is always taken to lie below the weld under consideration.
2. The inclination of the weld axis is measured from the horizontal reference plane toward the vertical reference plane.
3. The angle of rotation of the weld face is determined by a line perpendicular to the weld face at its center which passes through the weld axis. The reference position (0°) of rotation of the weld face invariably points in the direction opposite to that in which the weld axis angle increases. When looking at point P, the angle of rotation of the weld face is measured in a clockwise direction from the reference position (0°).

Figure 16B — Welding Position Diagram for Fillet Welds in Plate

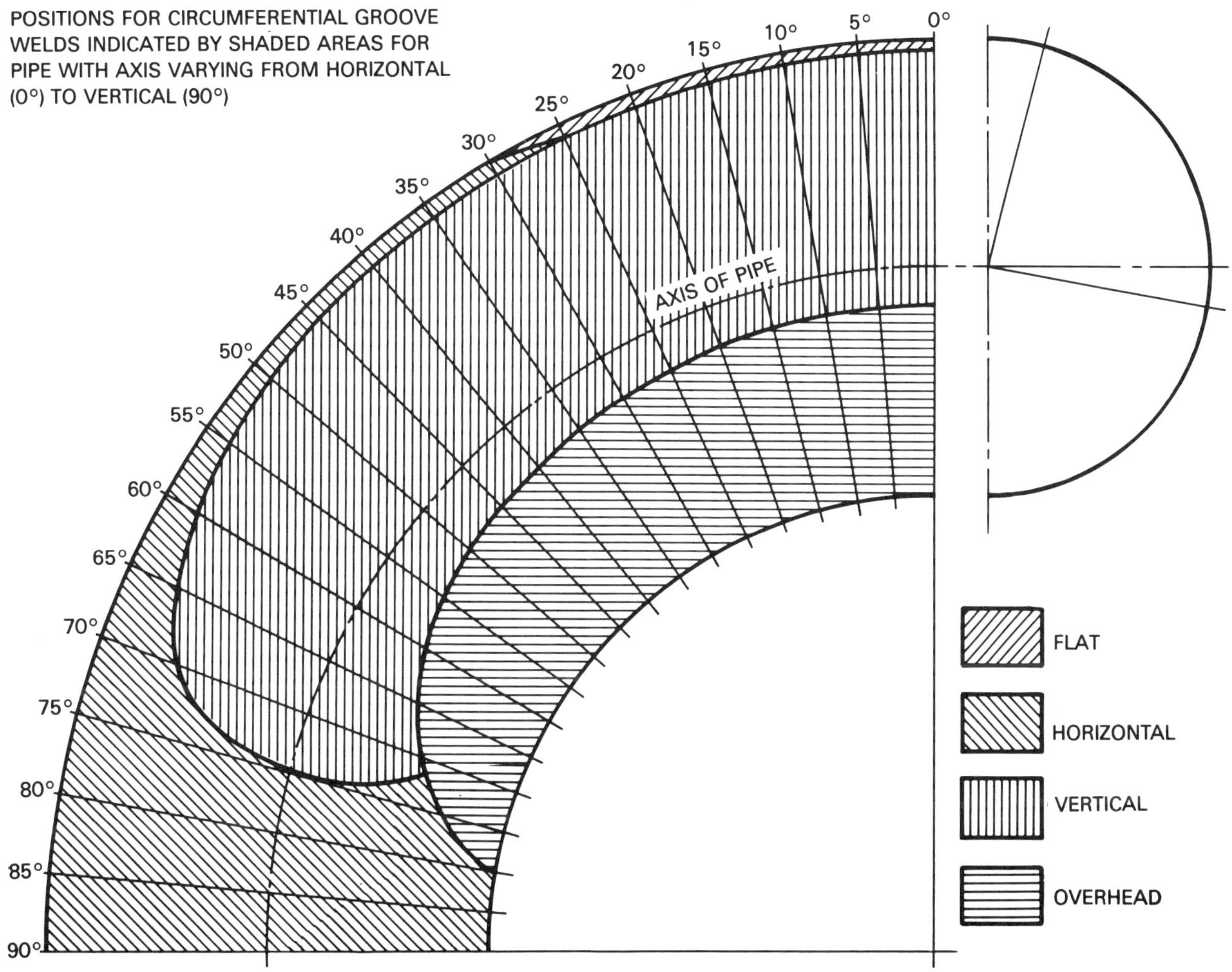

(F)

Figure 16C — Welding Position Diagram for Groove Welds in Pipe

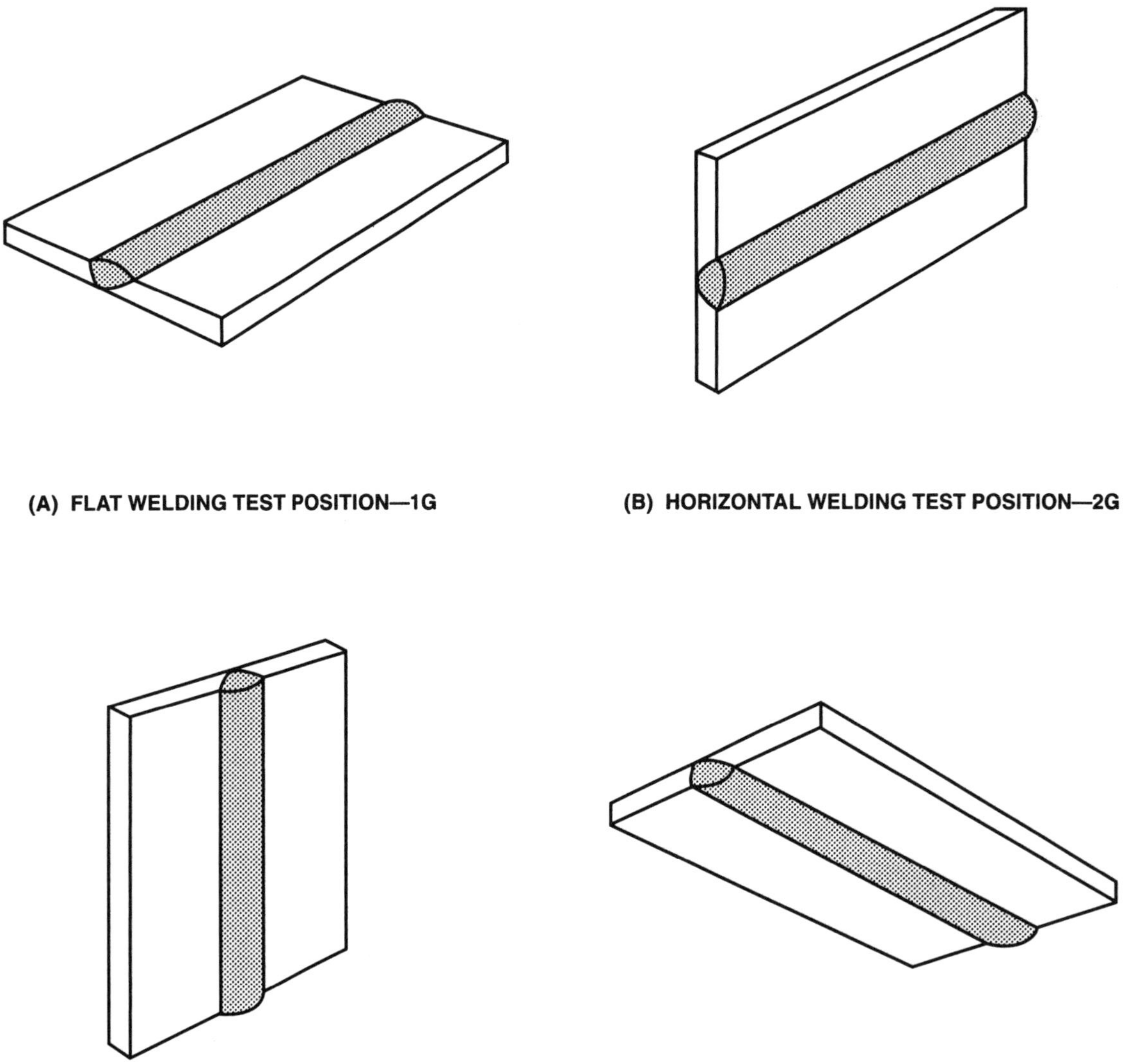

(A) FLAT WELDING TEST POSITION—1G

(B) HORIZONTAL WELDING TEST POSITION—2G

(C) VERTICAL WELDING TEST POSITION—3G

(D) OVERHEAD WELDING TEST POSITION—4G

Figure 17 — Welding Test Positions and Their Designations for Groove Welds in Plate

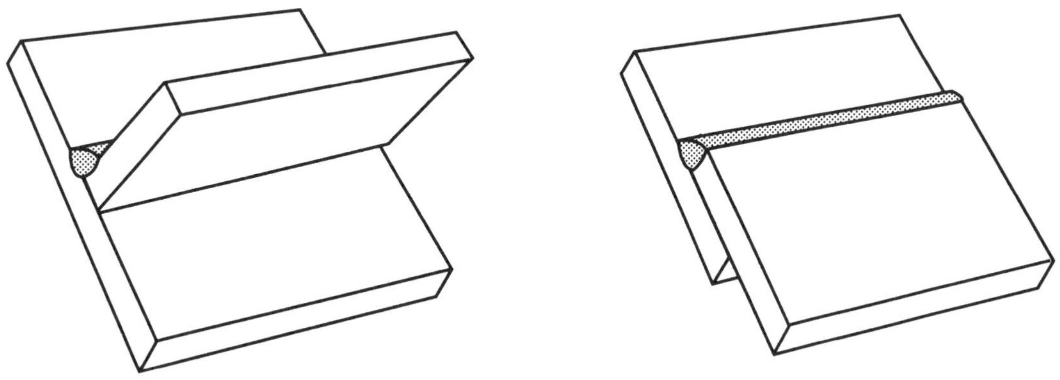

(A) FLAT WELDING TEST POSITION—1F

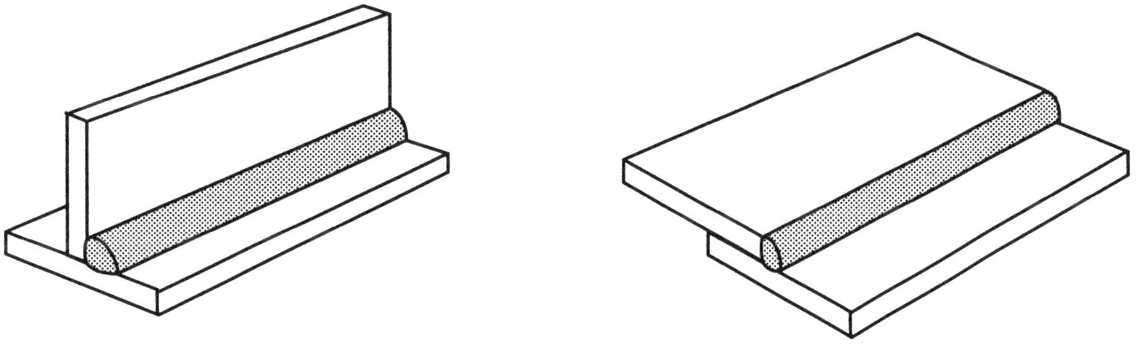

(B) HORIZONTAL WELDING TEST POSITION—2F

Figure 18 — Welding Test Positions and Their Designations for Fillet Welds in Plate

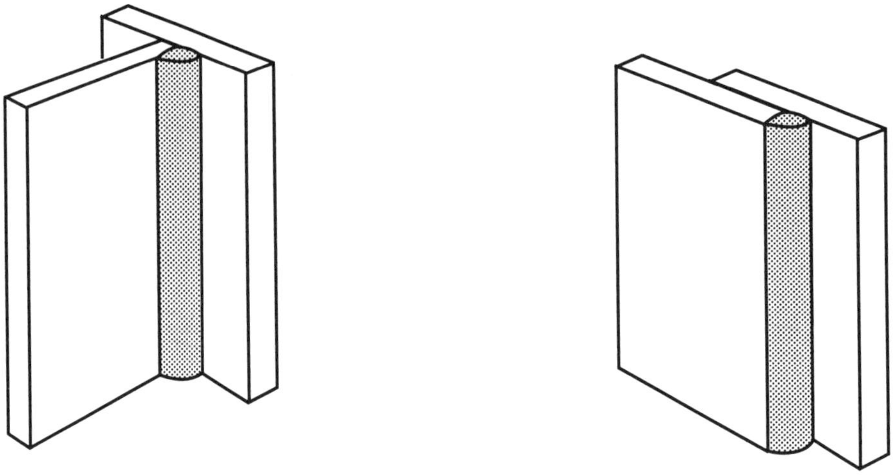

(C) VERTICAL WELDING TEST POSITION—3F

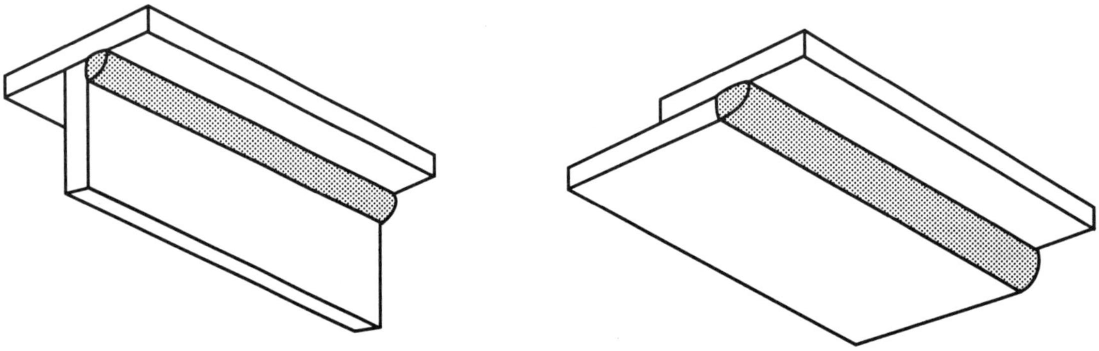

(D) OVERHEAD WELDING TEST POSITION—4F

Figure 18 (Continued) — Welding Test Positions and Their Designations for Fillet Welds in Plate

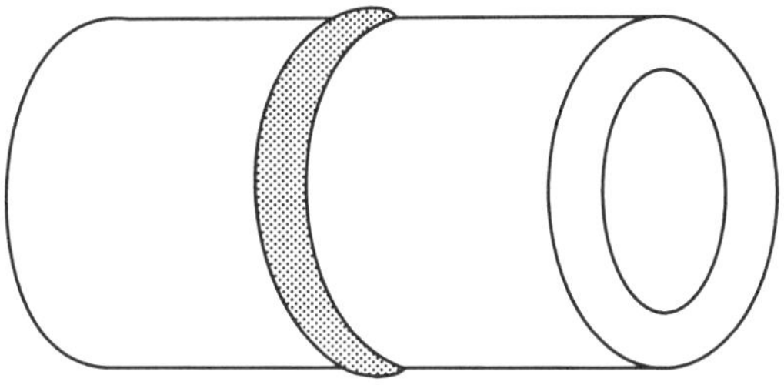

PIPE ROTATED

(A) FLAT WELDING TEST POSITION—1G

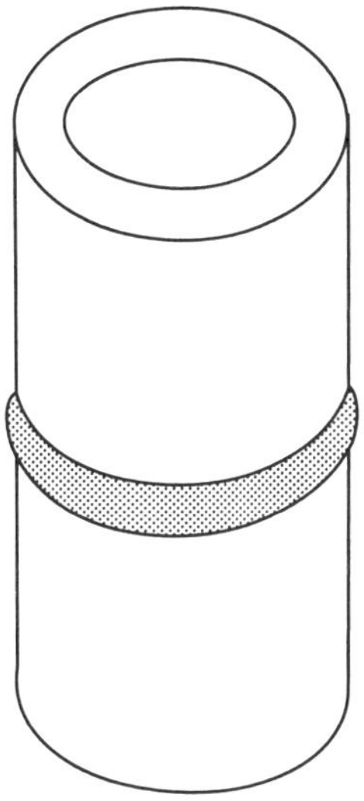

(B) HORIZONTAL WELDING TEST POSITION—2G

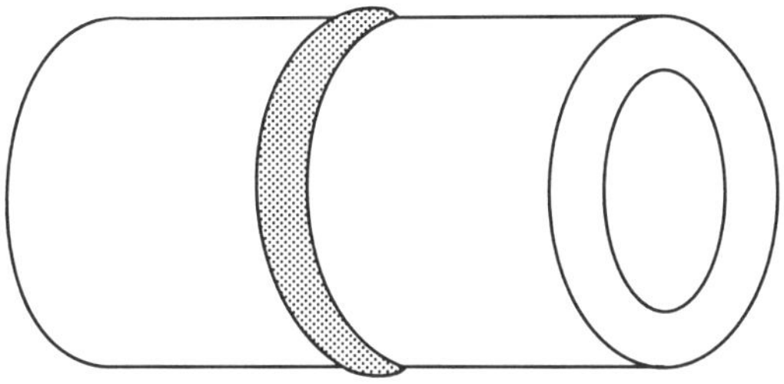

PIPE FIXED

(C) MULTIPLE WELDING TEST POSITION—5G

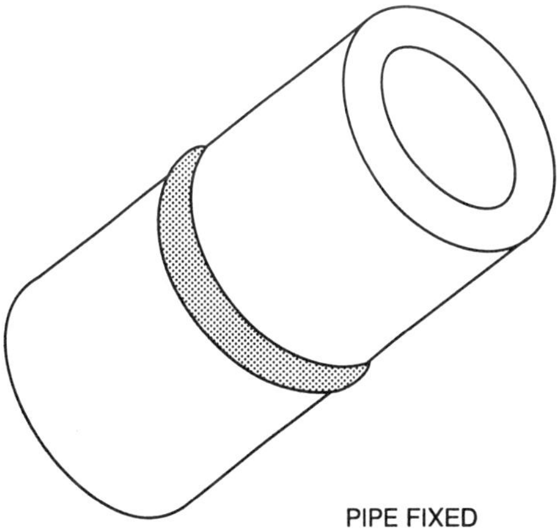

PIPE FIXED

(D) MULTIPLE WELDING TEST POSITION—6GR

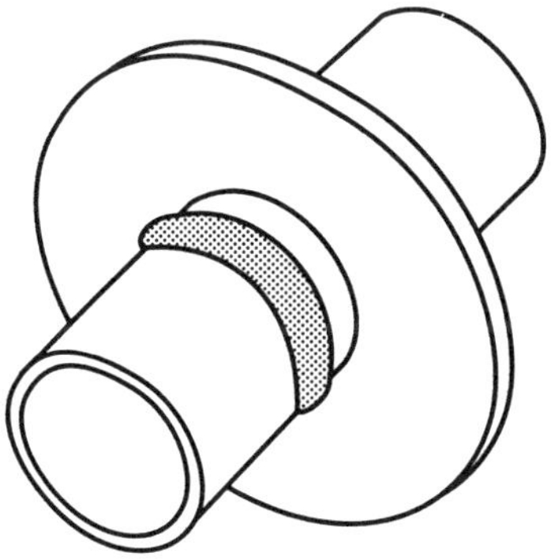

(E) MULTIPLE WELDING TEST POSITION WITH RESTRICTION RING—6GR

Figure 19 — Welding Test Positions and Their Designations for Groove Welds in Pipe

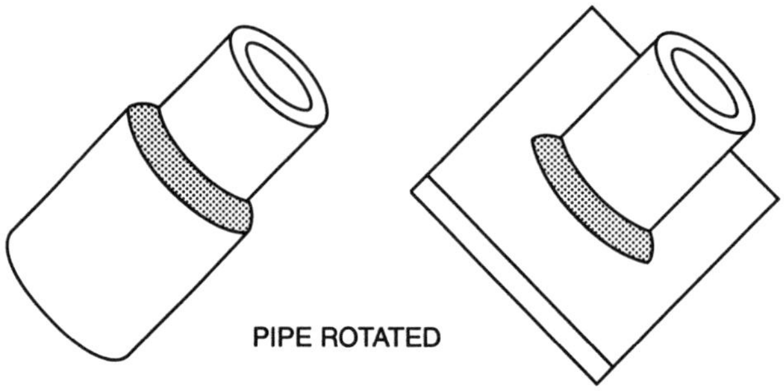

(A) FLAT WELDING TEST POSITION—1F

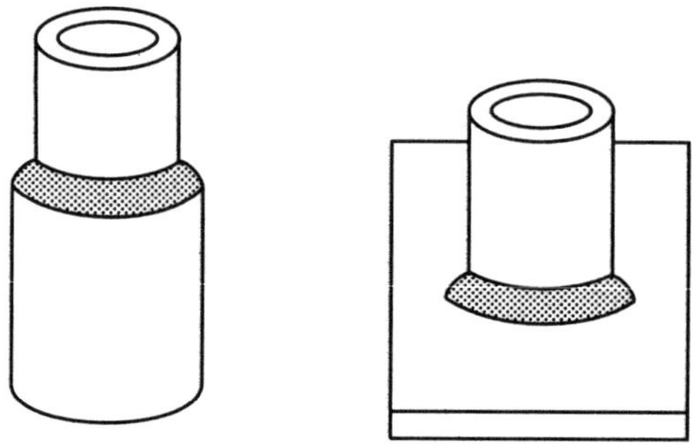

(B) HORIZONTAL WELDING TEST POSITION—2F

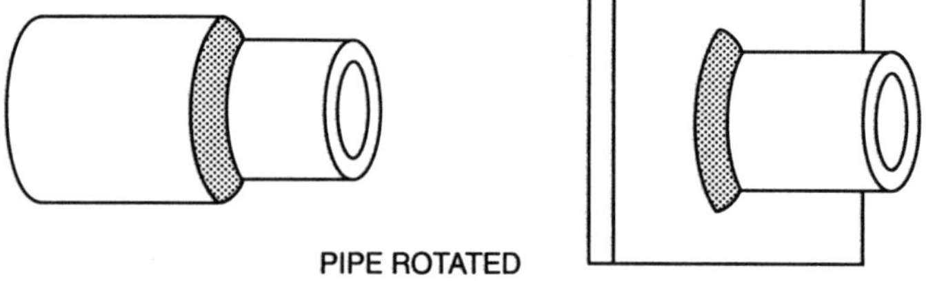

(C) HORIZONTAL WELDING TEST POSITION—2FR

Figure 20 — Welding Test Positions and Their Designations for Fillet Welds in Pipe

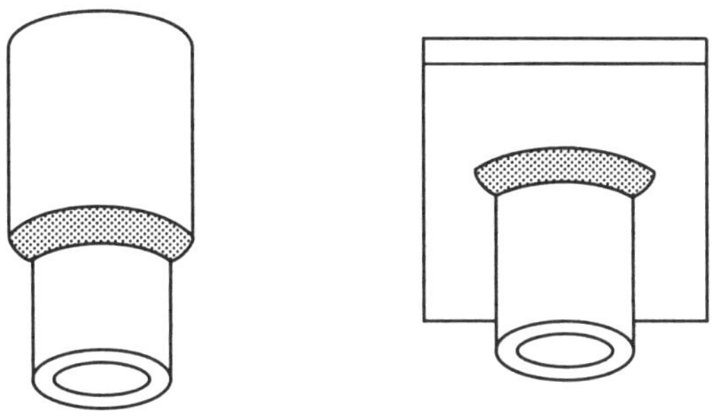

(D) OVERHEAD WELDING TEST POSITION—4F

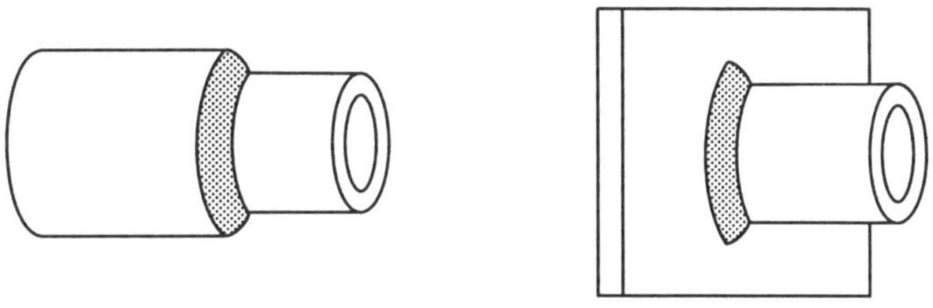

(E) MULTIPLE WELDING TEST POSITION—5F

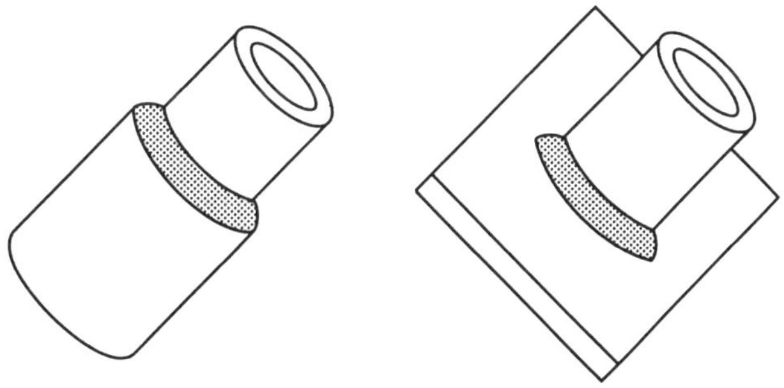

(F) MULTIPLE WELDING TEST POSITION—6F

Figure 20 (Continued)— Welding Test Positions and Their Designations for Fillet Welds in Pipe

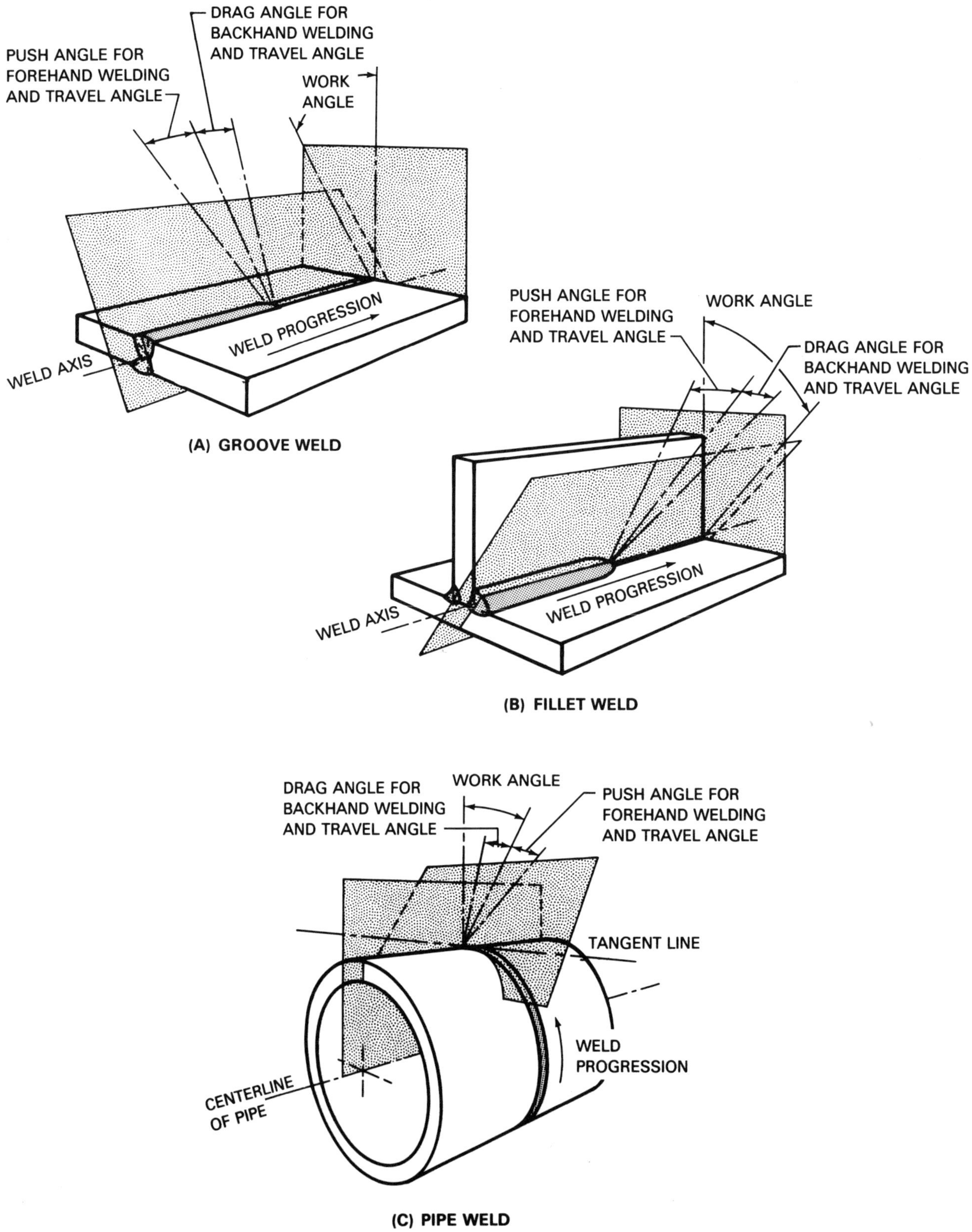

Figure 21 — Position of Electrode, Gun, Torch, Rod, or Beam

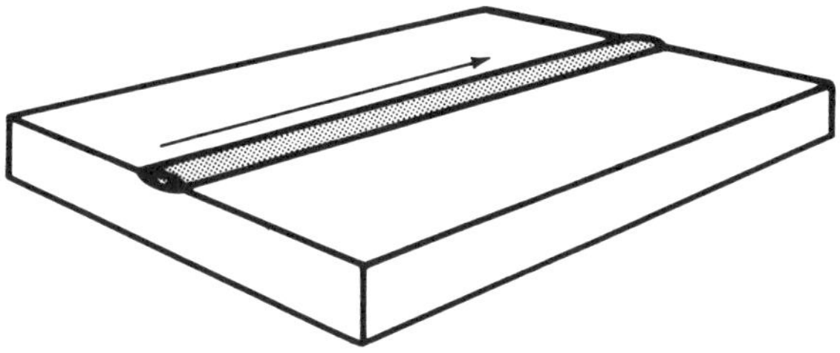

(A) STRINGER BEAD

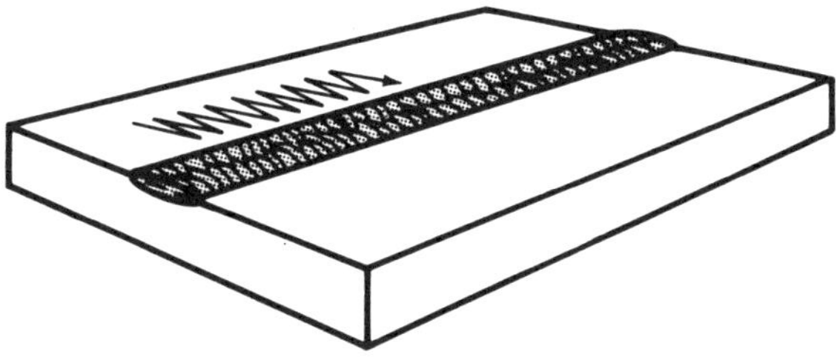

(B) WEAVE BEAD

Note: The arrows adjacent to the weld beads indicate the relative motion of the beam, electrode, or flame relative to the workpiece.

Figure 22 — Weld Bead Types

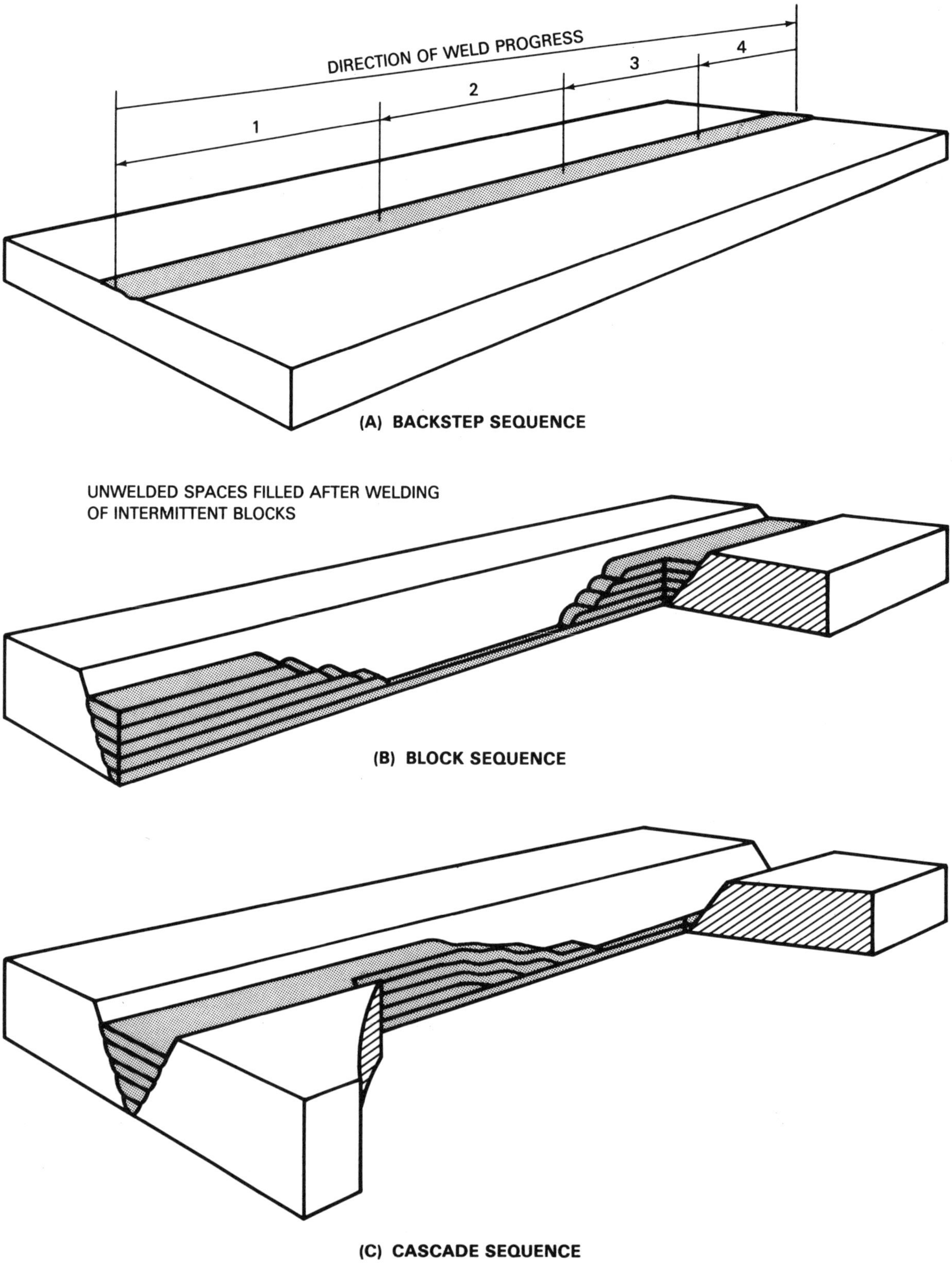

Figure 23 — Welding Sequence

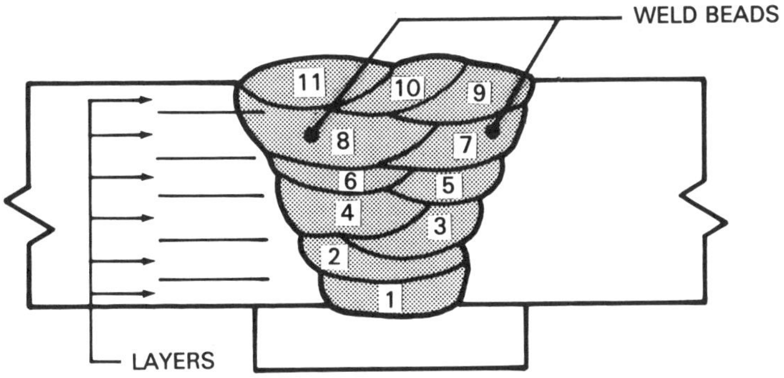

(D) CROSS-SECTIONAL SEQUENCE

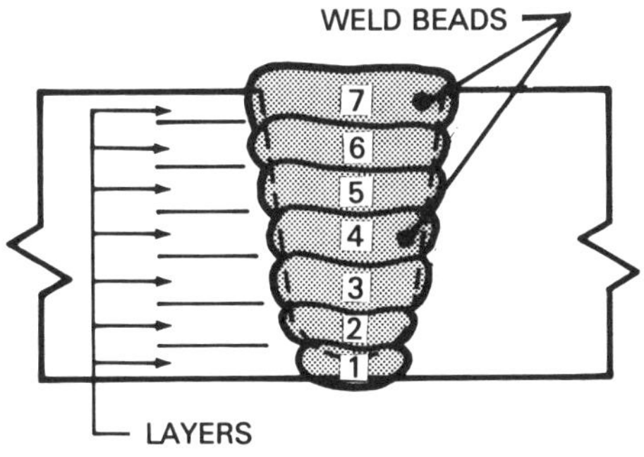

(E) CROSS-SECTIONAL SEQUENCE

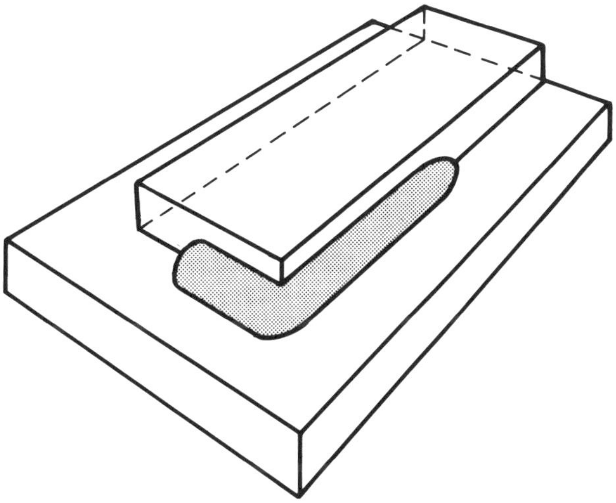

(F) BOXING

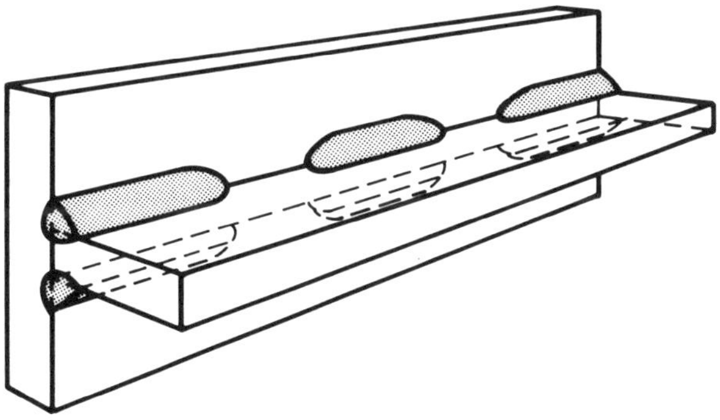

(G) CHAIN INTERMITTENT FILLET WELD

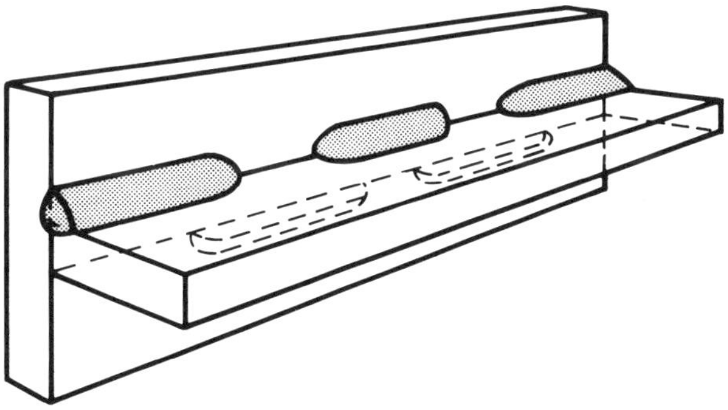

(H) STAGGERED INTERMITTENT FILLET WELD

Figure 23 (Continued) — Welding Sequence

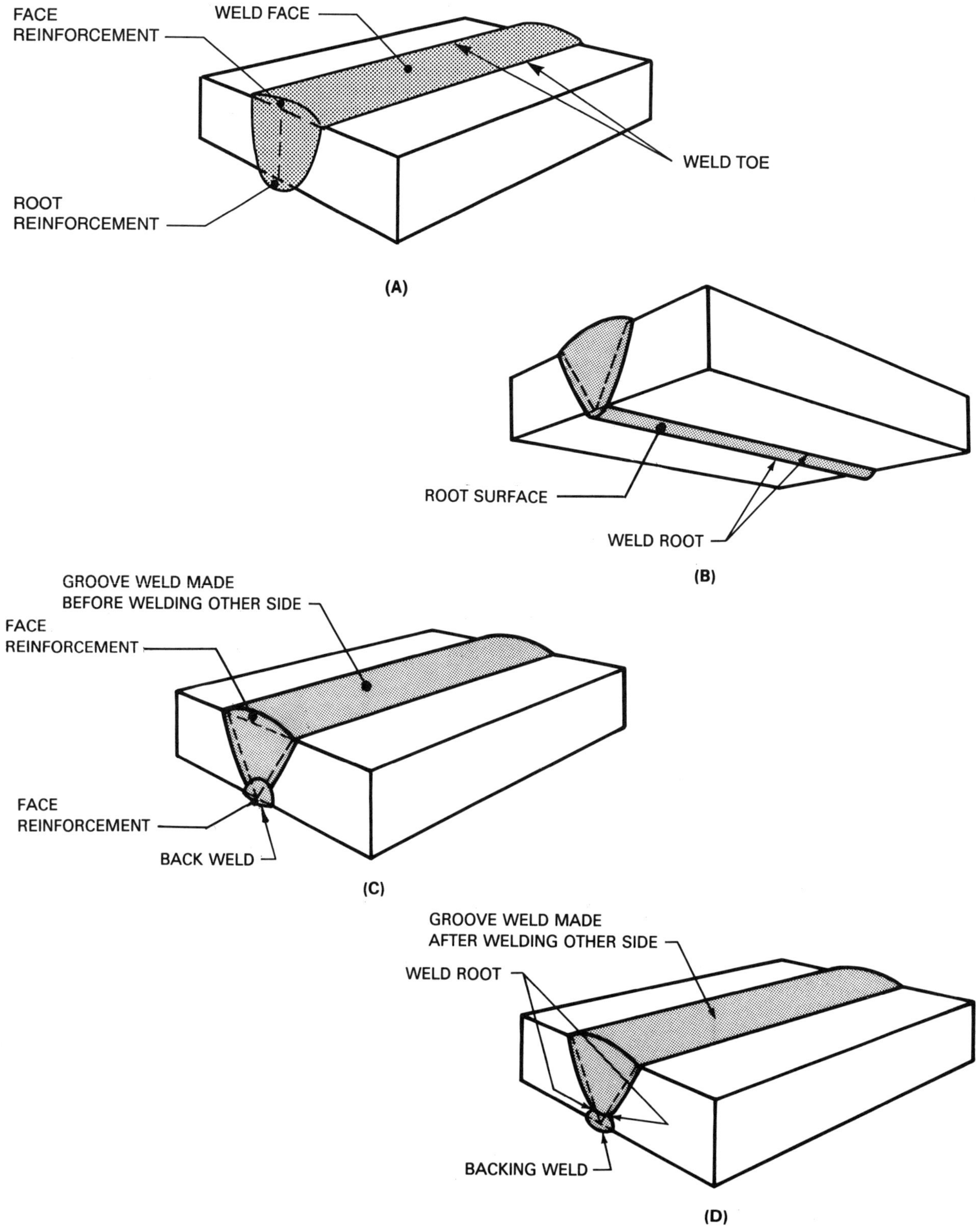

Figure 24 — Parts of a Weld

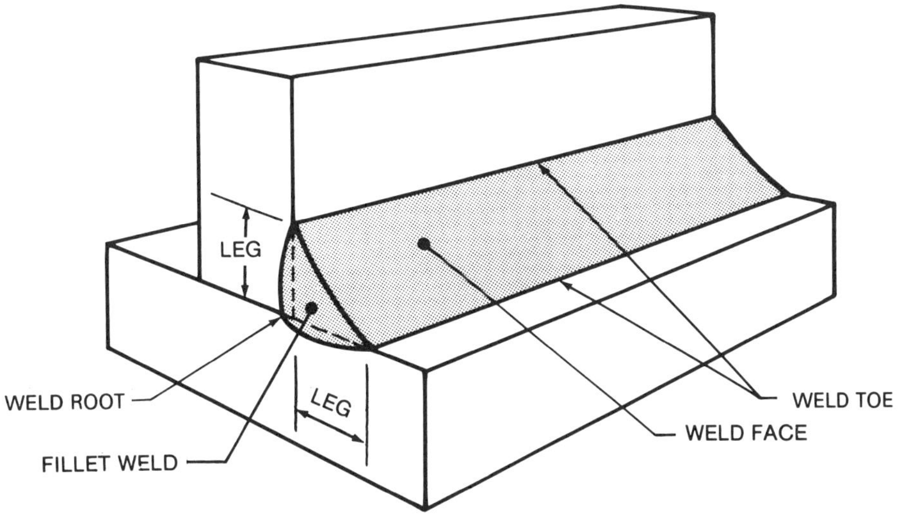

(E)

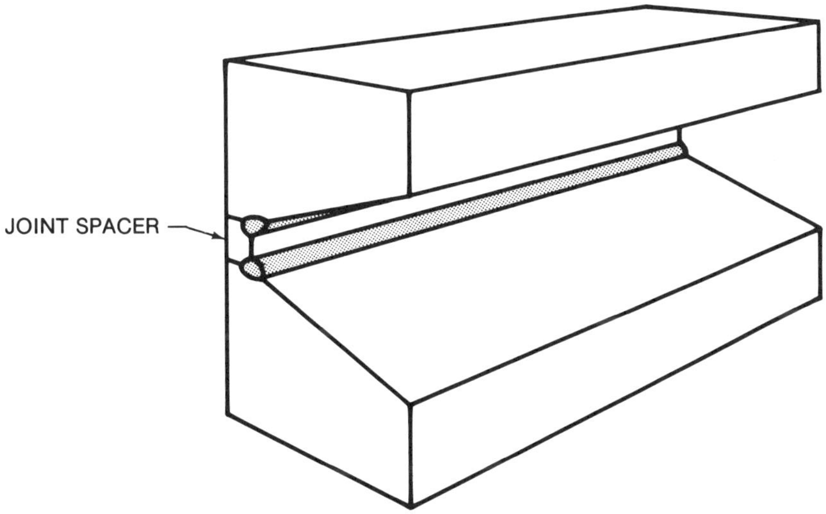

(F)

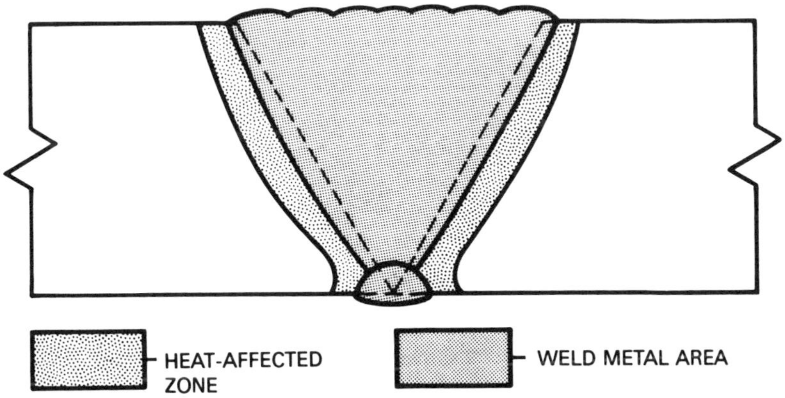

(G)

Figure 24 (Continued) — Parts of a Weld

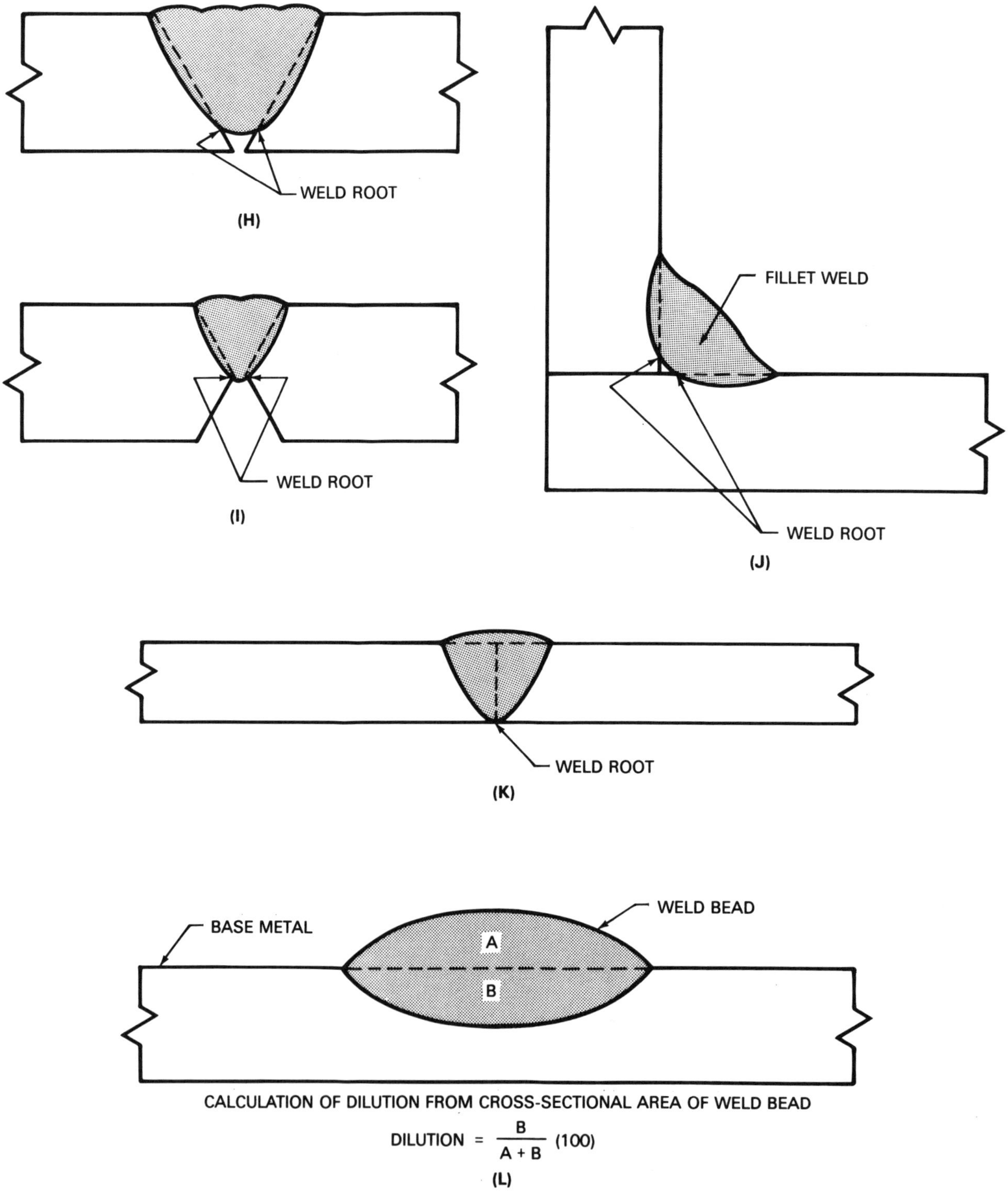

Figure 24 (Continued) — Parts of a Weld

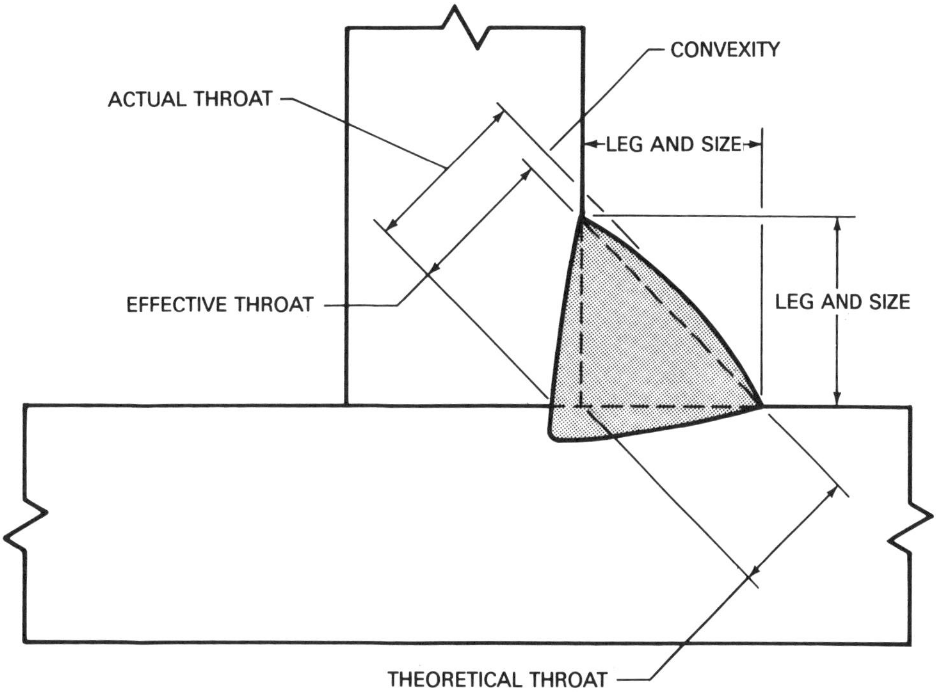

(A) CONVEX FILLET WELD

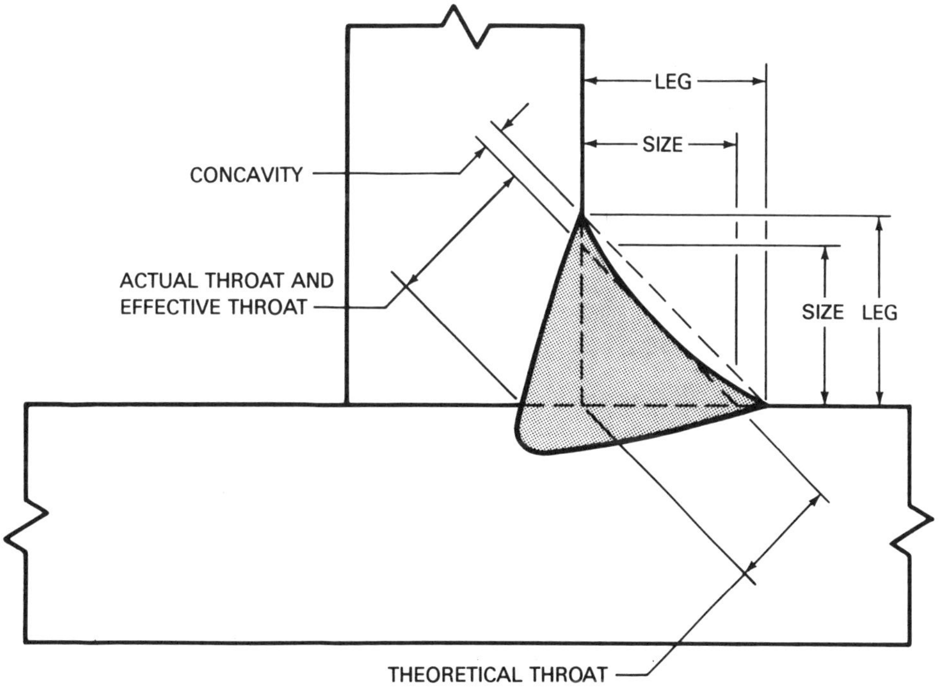

(B) CONCAVE FILLET WELD

Figure 25 — Weld Sizes

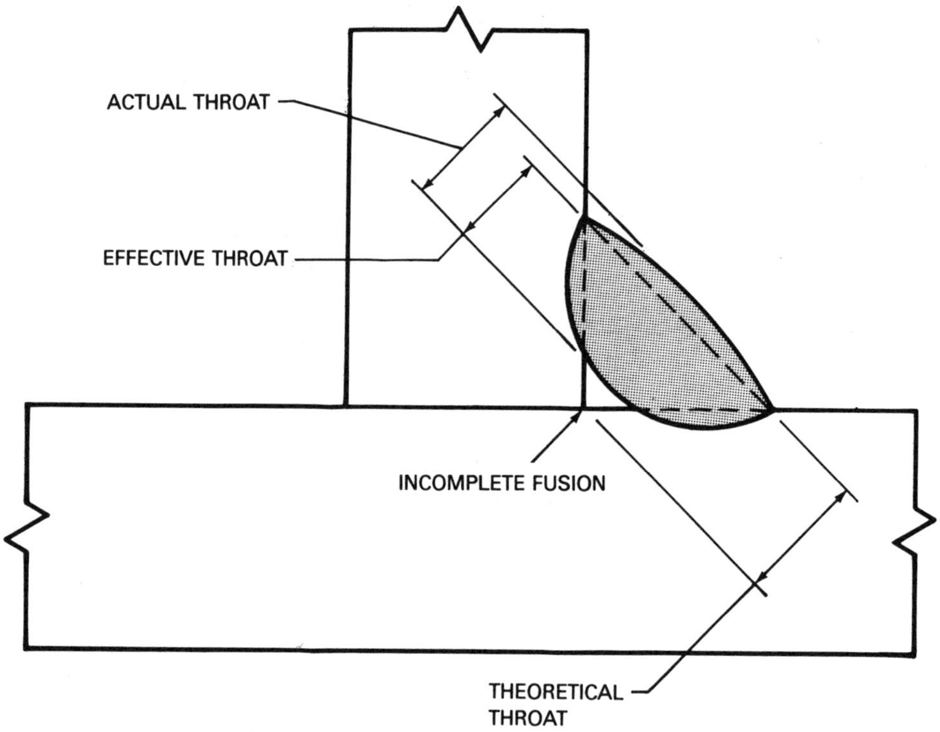

(C) FILLET WELD WITH INCOMPLETE FUSION

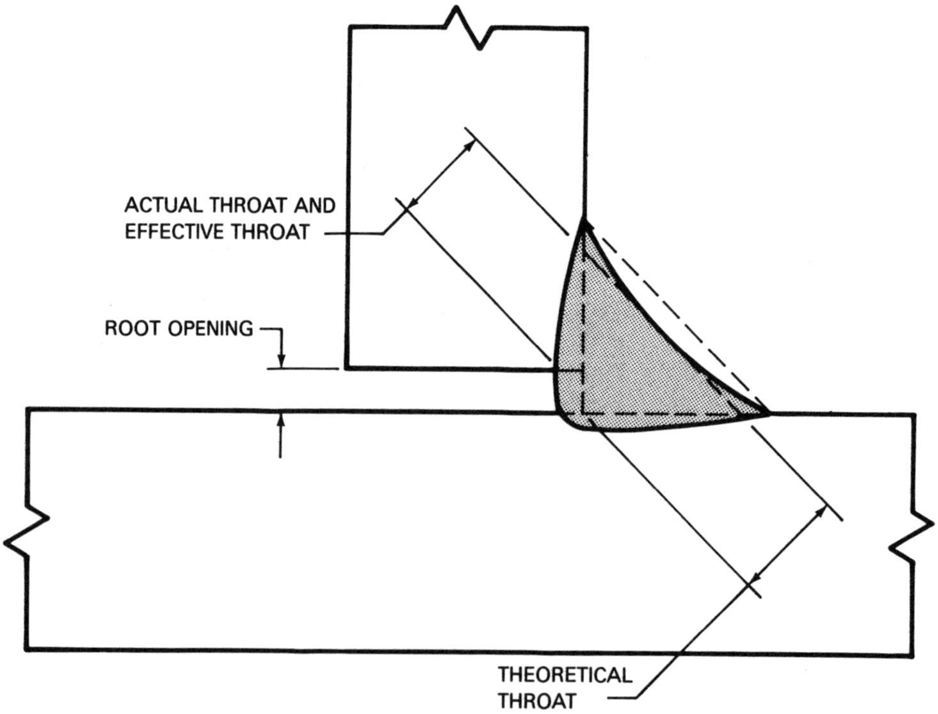

(D) T-JOINT WITH ROOT OPENING

Figure 25 (Continued) — Weld Sizes

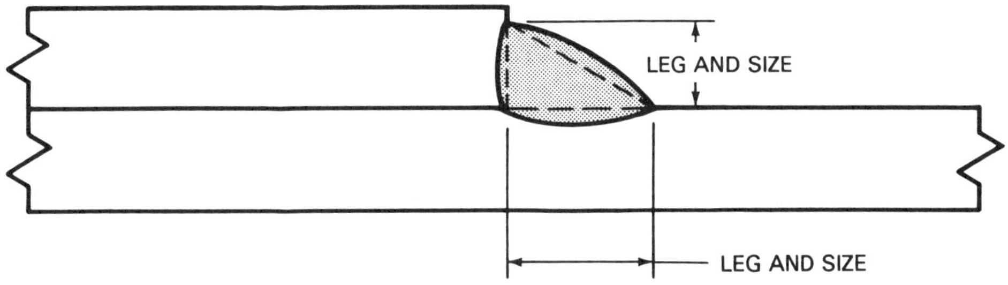

(E) UNEQUAL LEG FILLET WELD

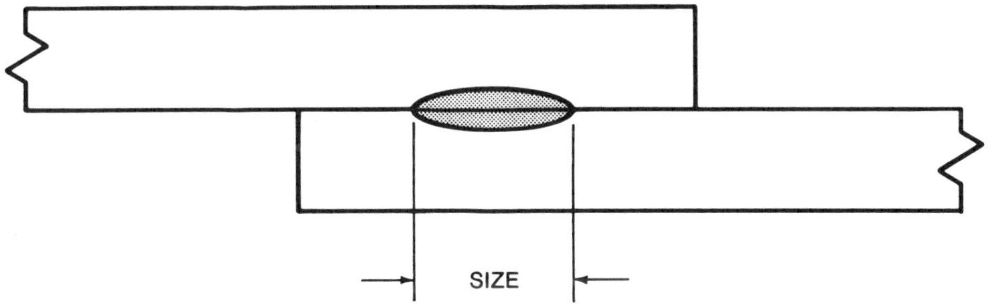

(F) SIZE OF SEAM OR SPOT WELD

Figure 25 (Continued) — Weld Sizes

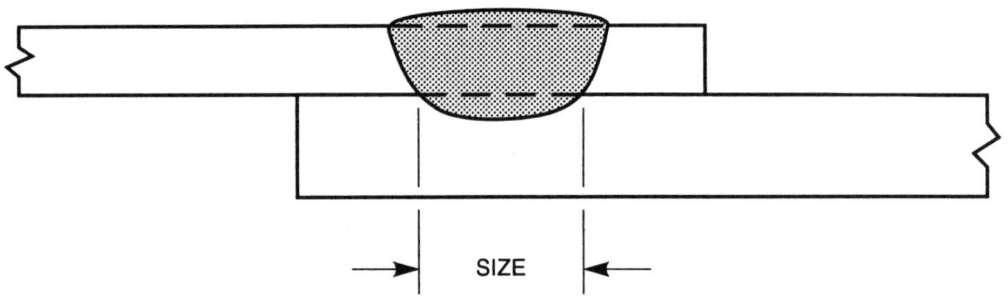

(G) ARC SEAM OR ARC SPOT WELD SIZE

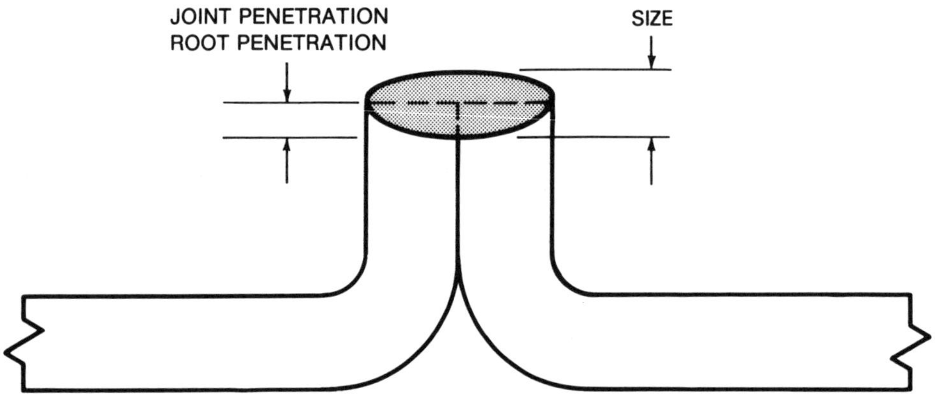

(H) EDGE WELD SIZE

Figure 25 (Continued) — Weld Sizes

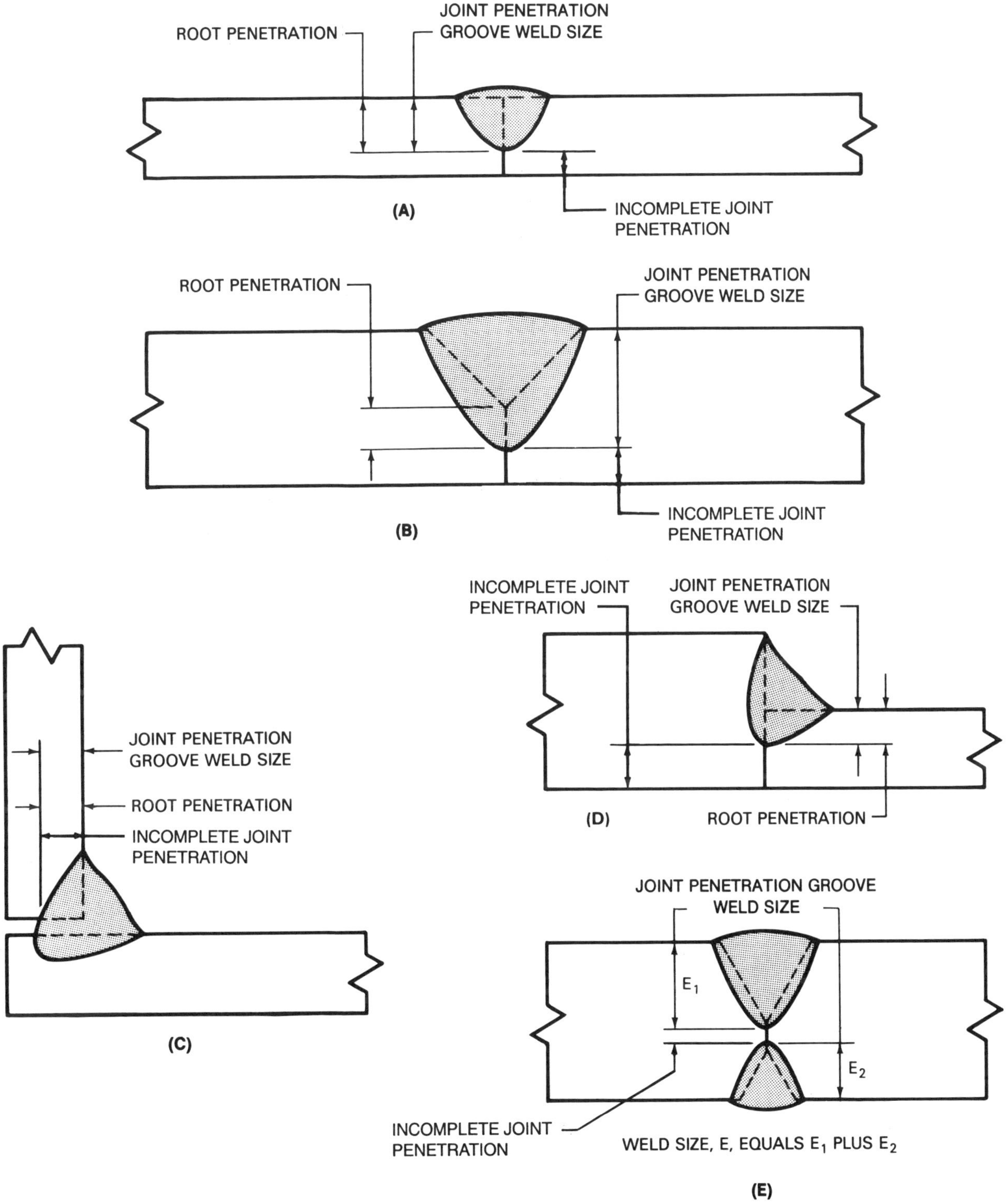

Figure 26 — Joint Penetration, Root Penetration, and Incomplete Joint Penetration

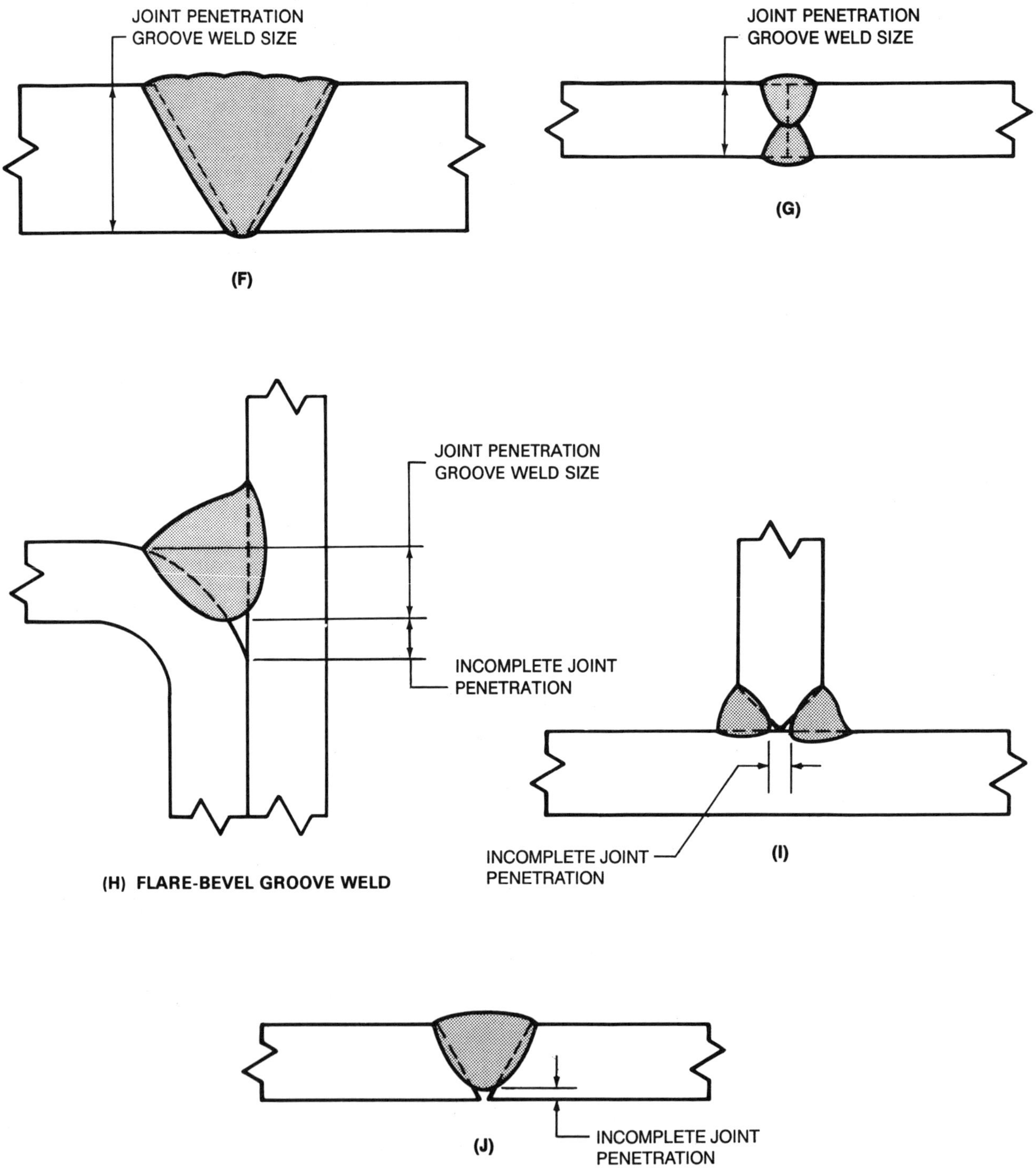

Figure 26 (Continued)—Joint Penetration, Root Penetration, and Incomplete Joint Penetration

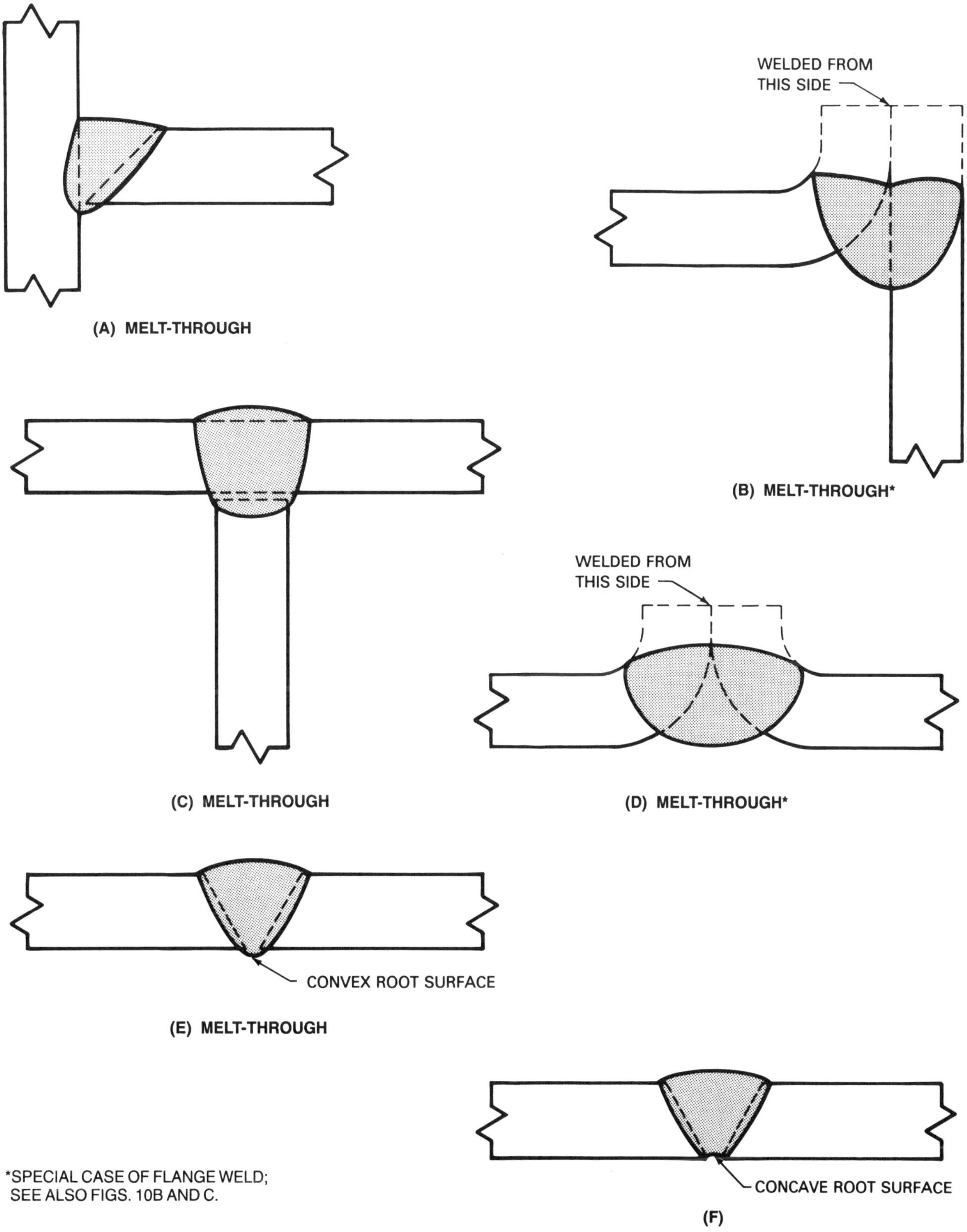

Figure 27 — Melt-Through and Root Surface Profile

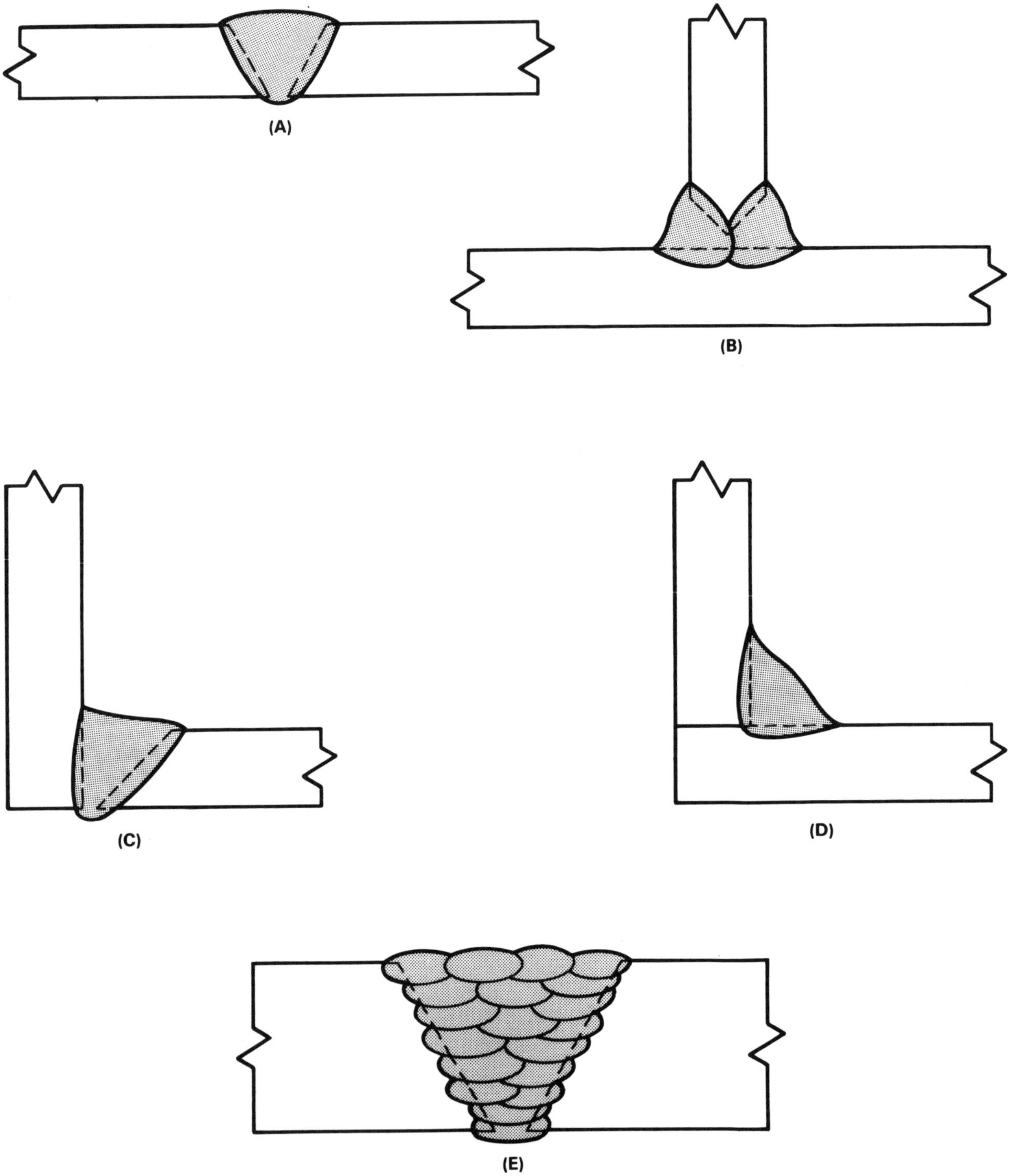

Figure 28 — Complete Fusion

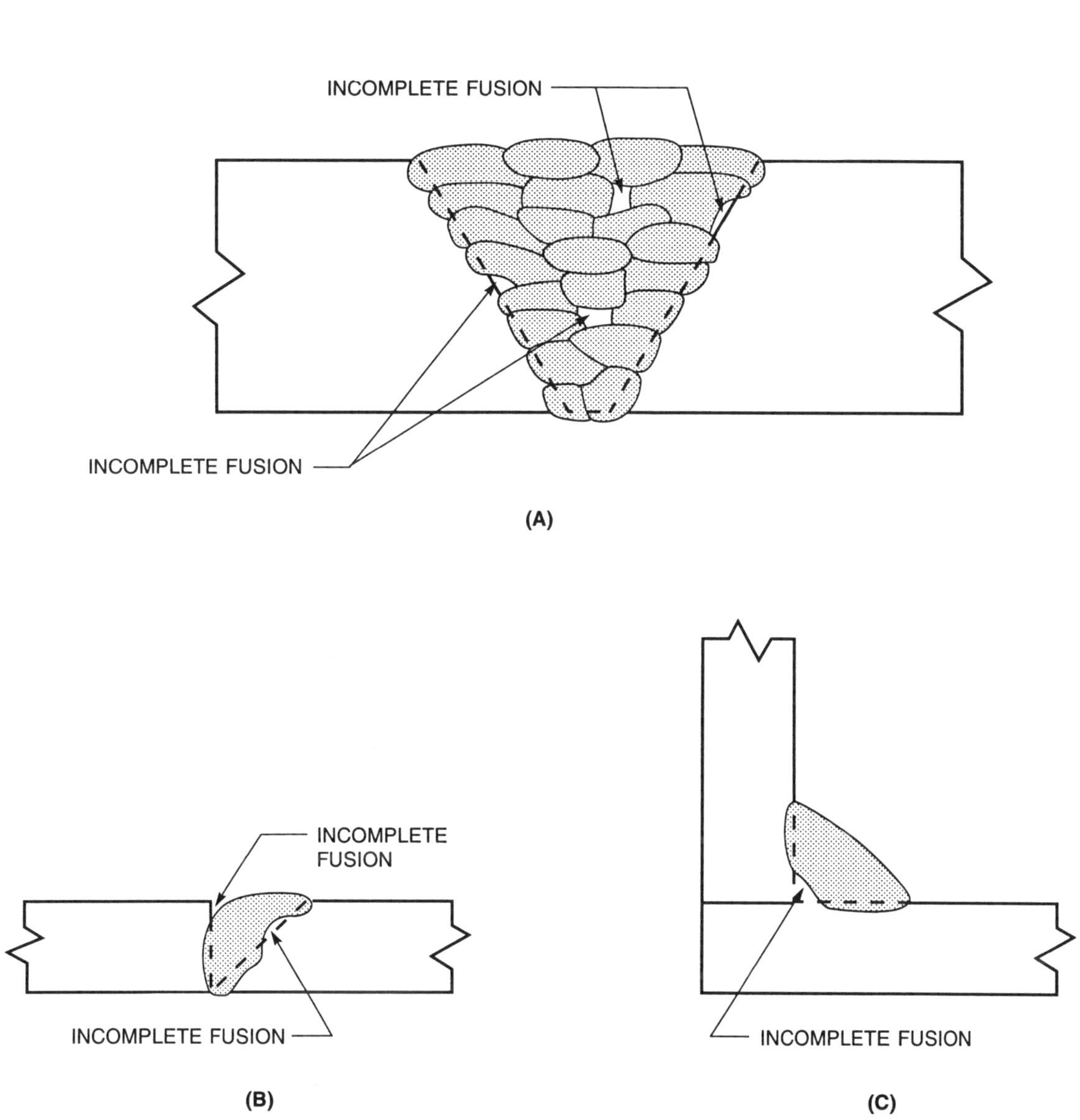

Figure 29 — Incomplete Fusion

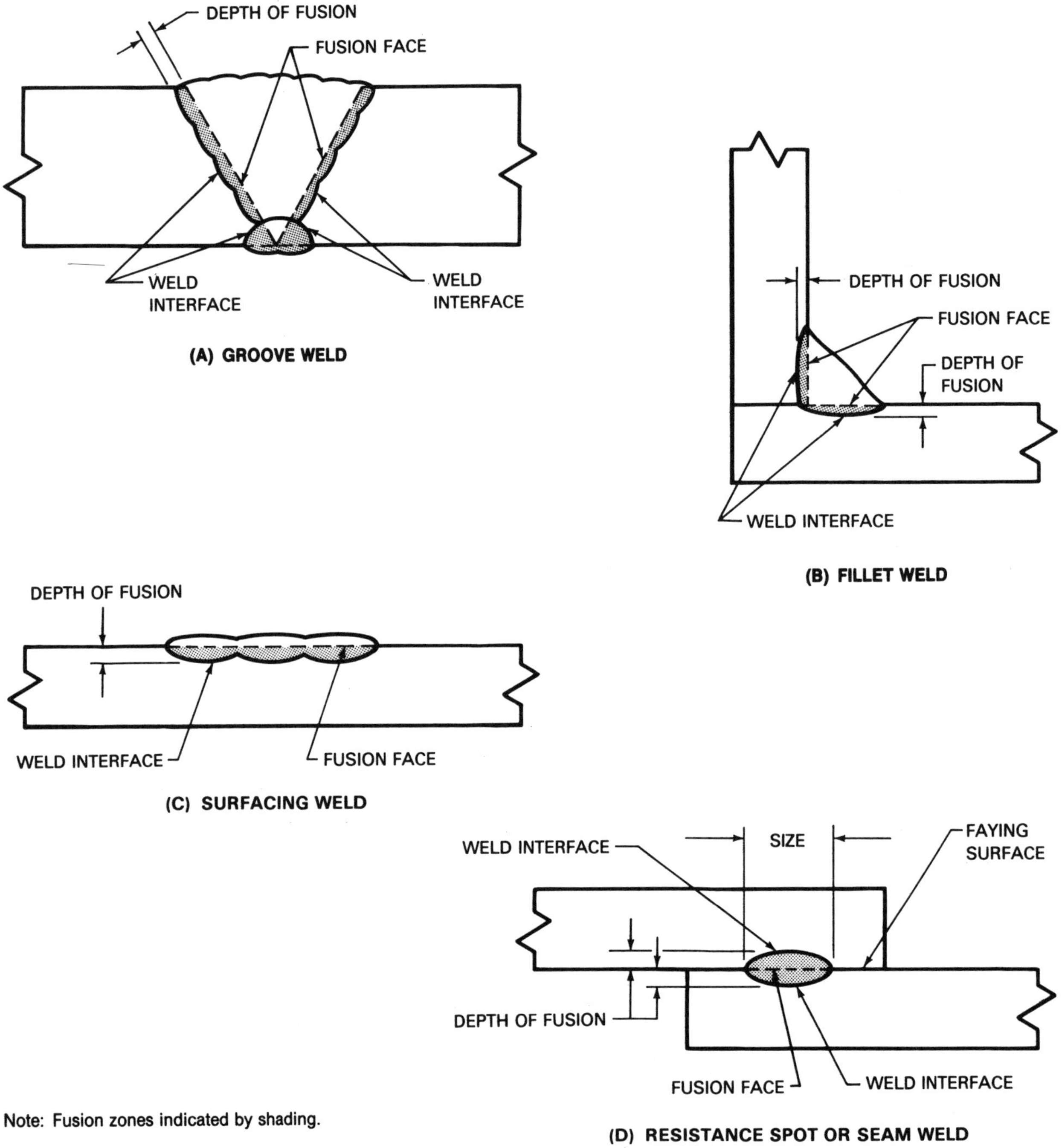

Figure 30 — Fusion Welds

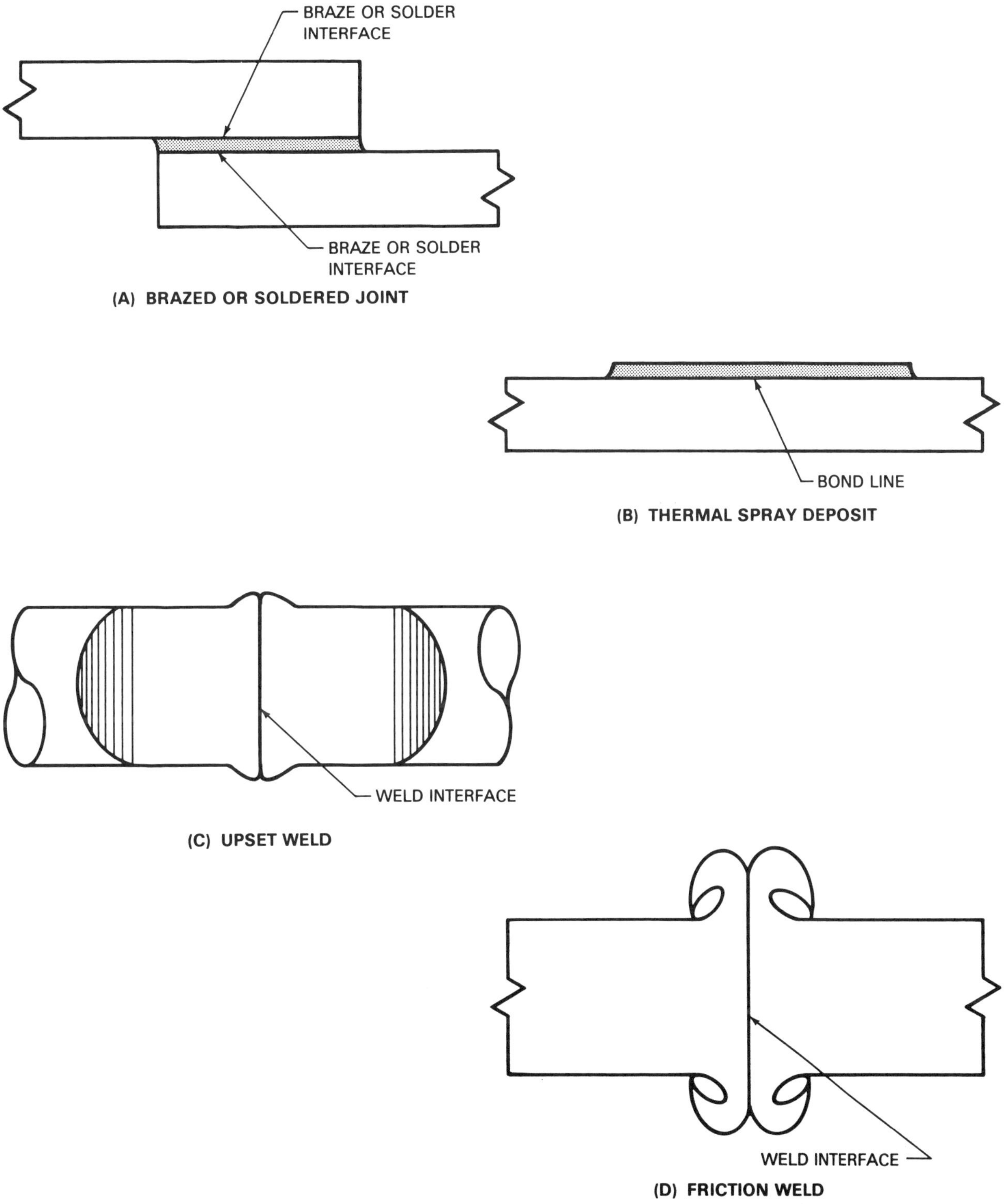

Figure 31 — Nonfusion Welds

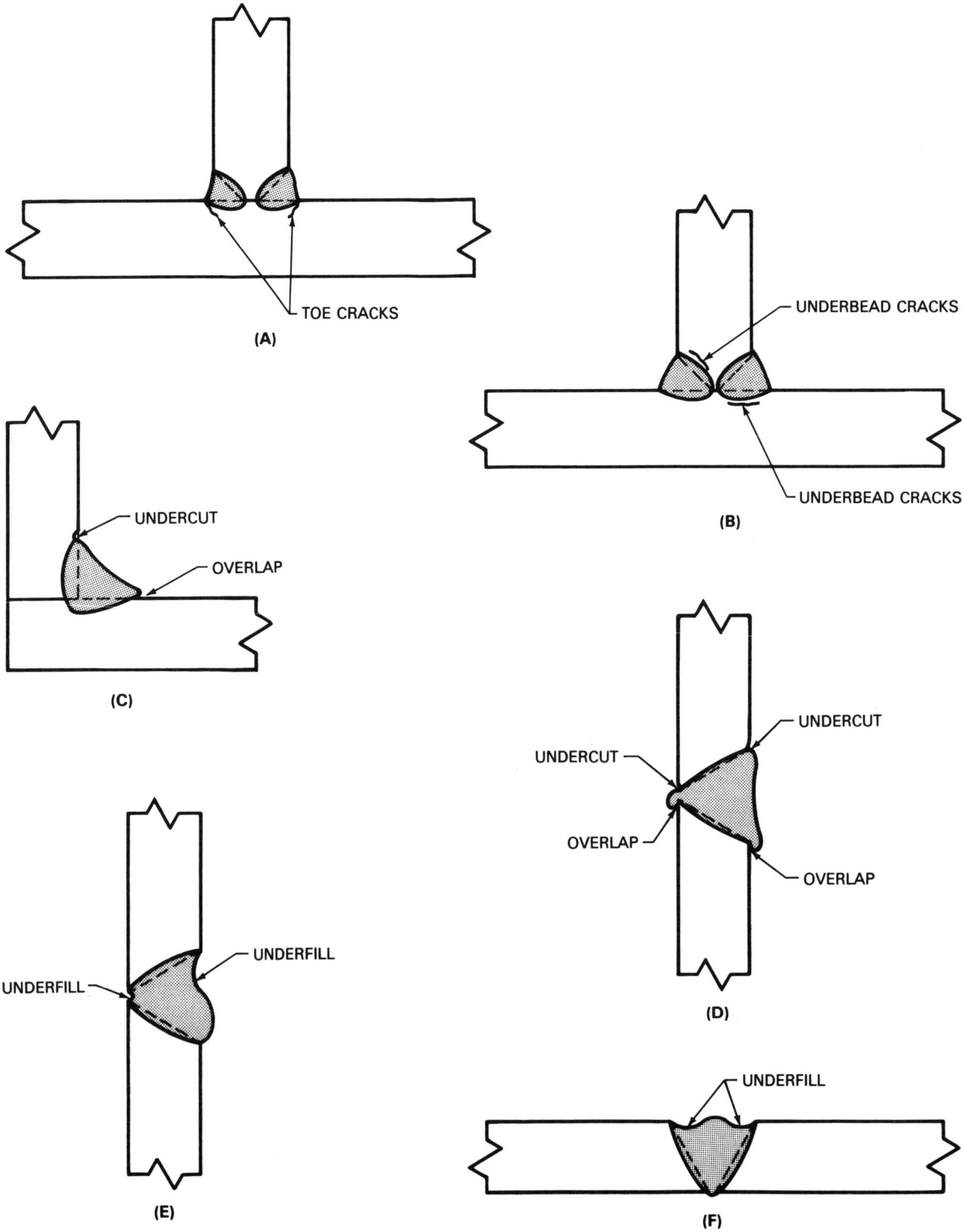

Figure 32 — Weld Discontinuities

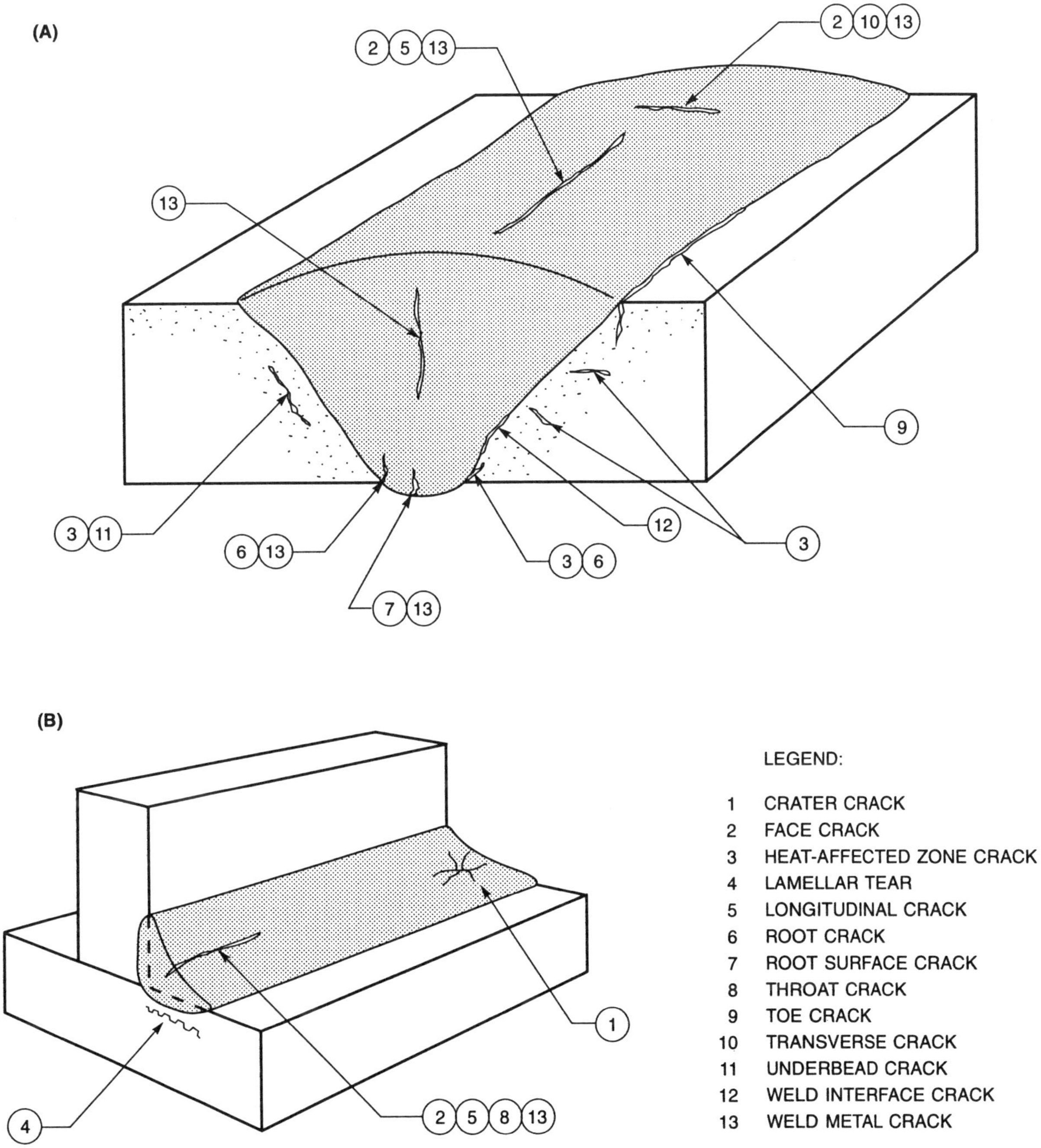

Figure 33 — Crack Types

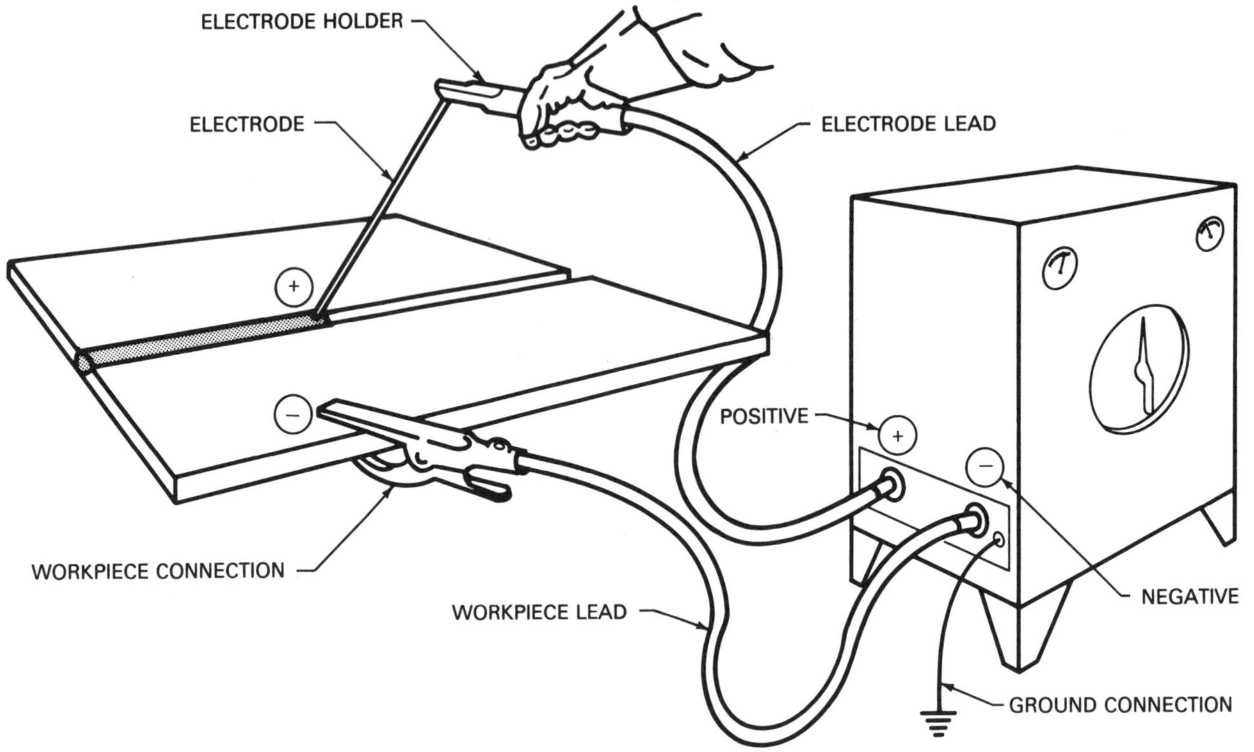

(A) DIRECT CURRENT ELECTRODE POSITIVE

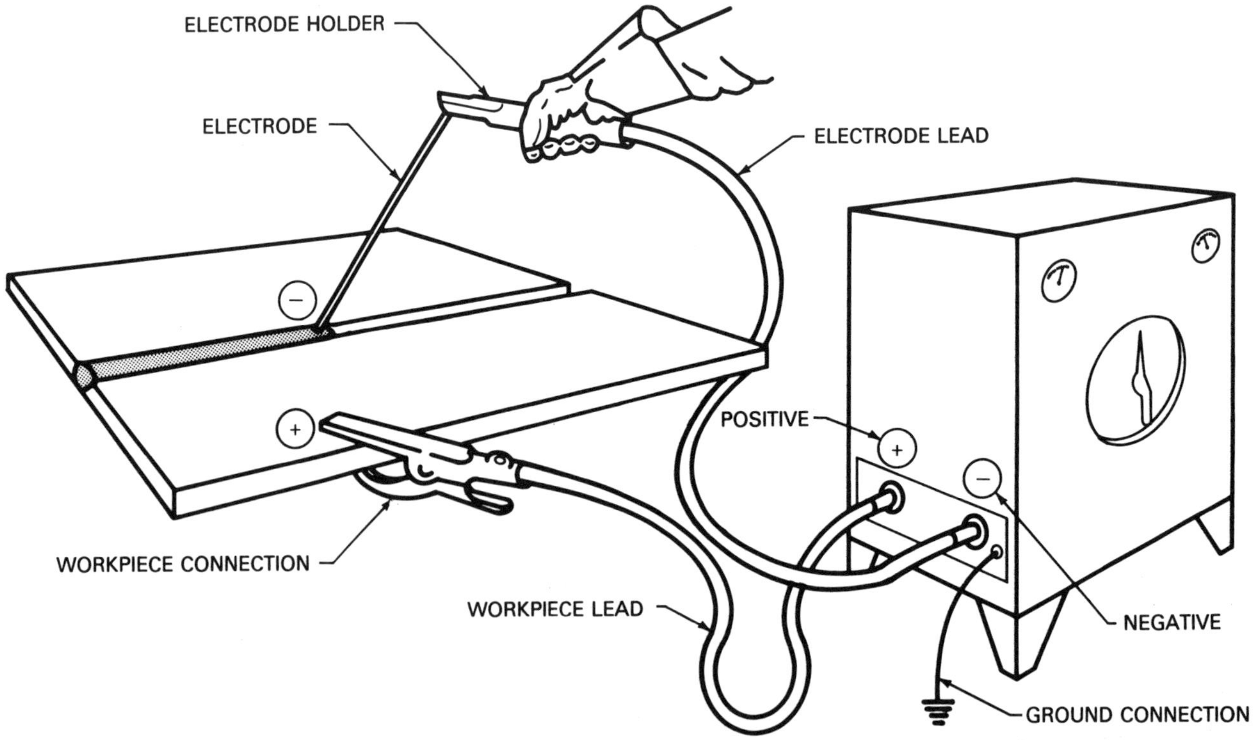

(B) DIRECT CURRENT ELECTRODE NEGATIVE

Figure 34 — Welding Current Polarity

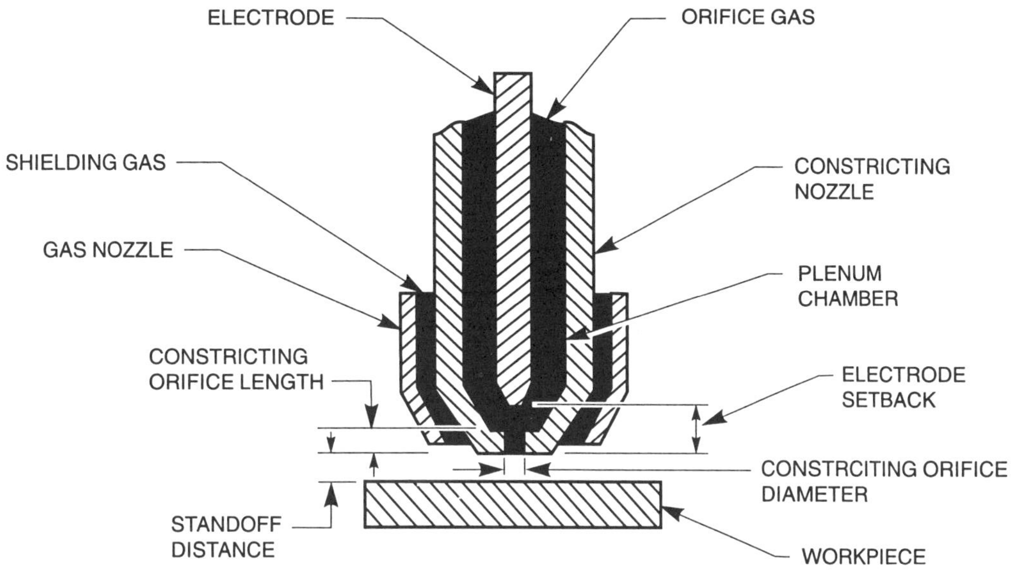

Figure 35 — Plasma Arc Torch Nomenclature

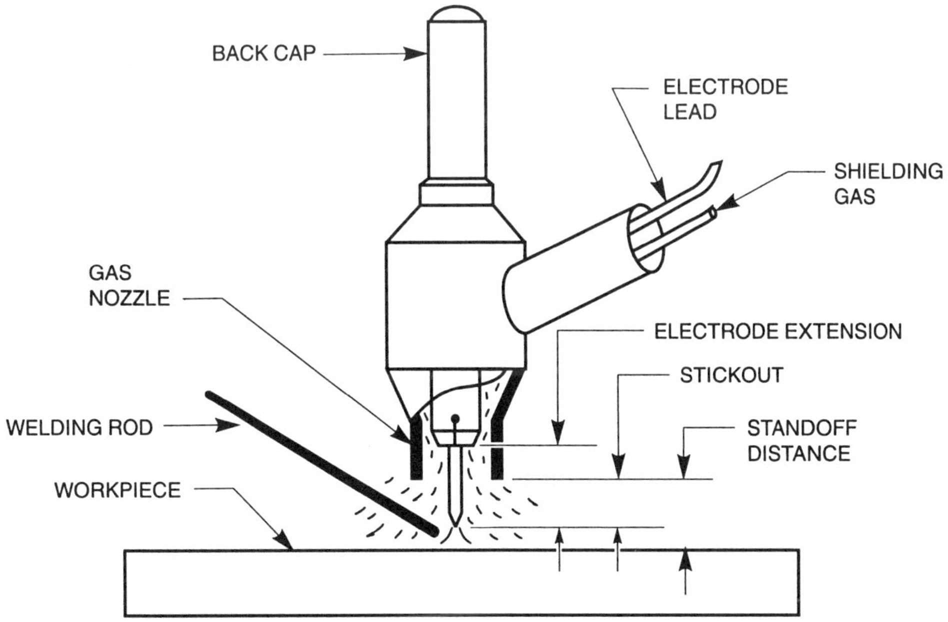

Figure 36 — Gas Tungsten Arc Welding Torch Nomenclature

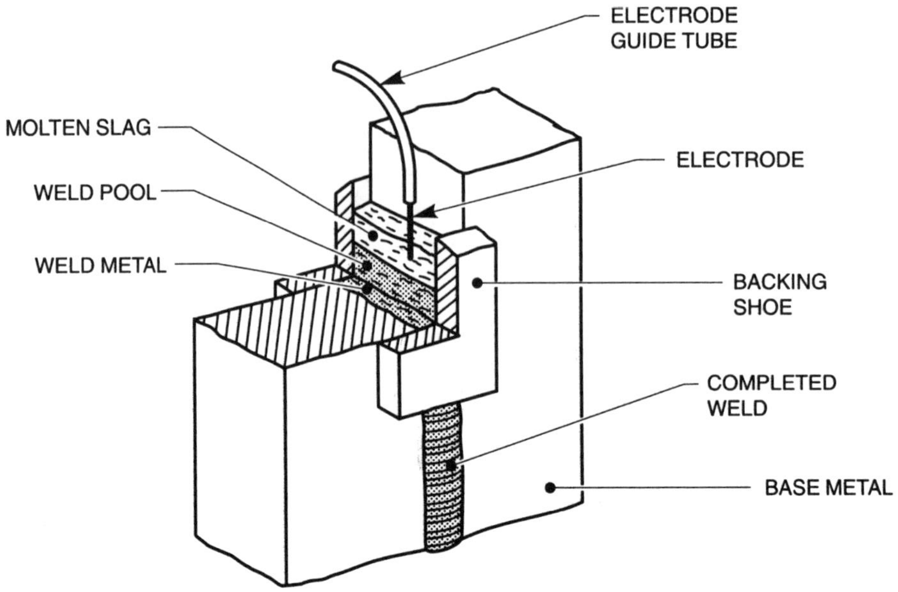

Figure 37 — Typical Arrangement for Electroslag Welding Process Nomenclature

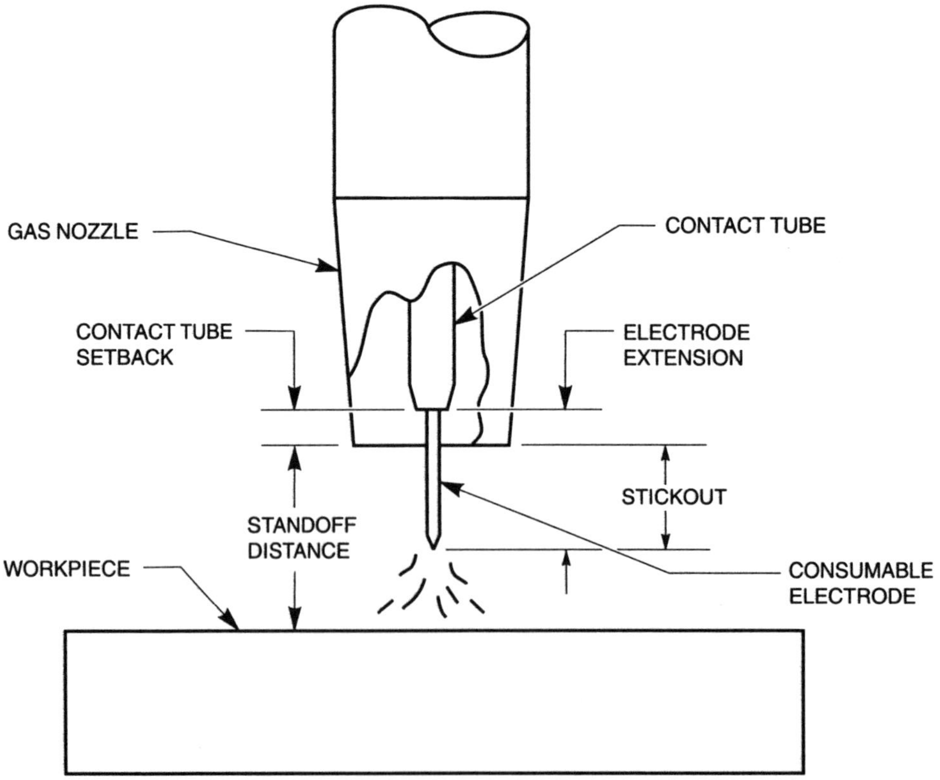

Figure 38 — Gas Metal Arc Welding Gun Nomenclature

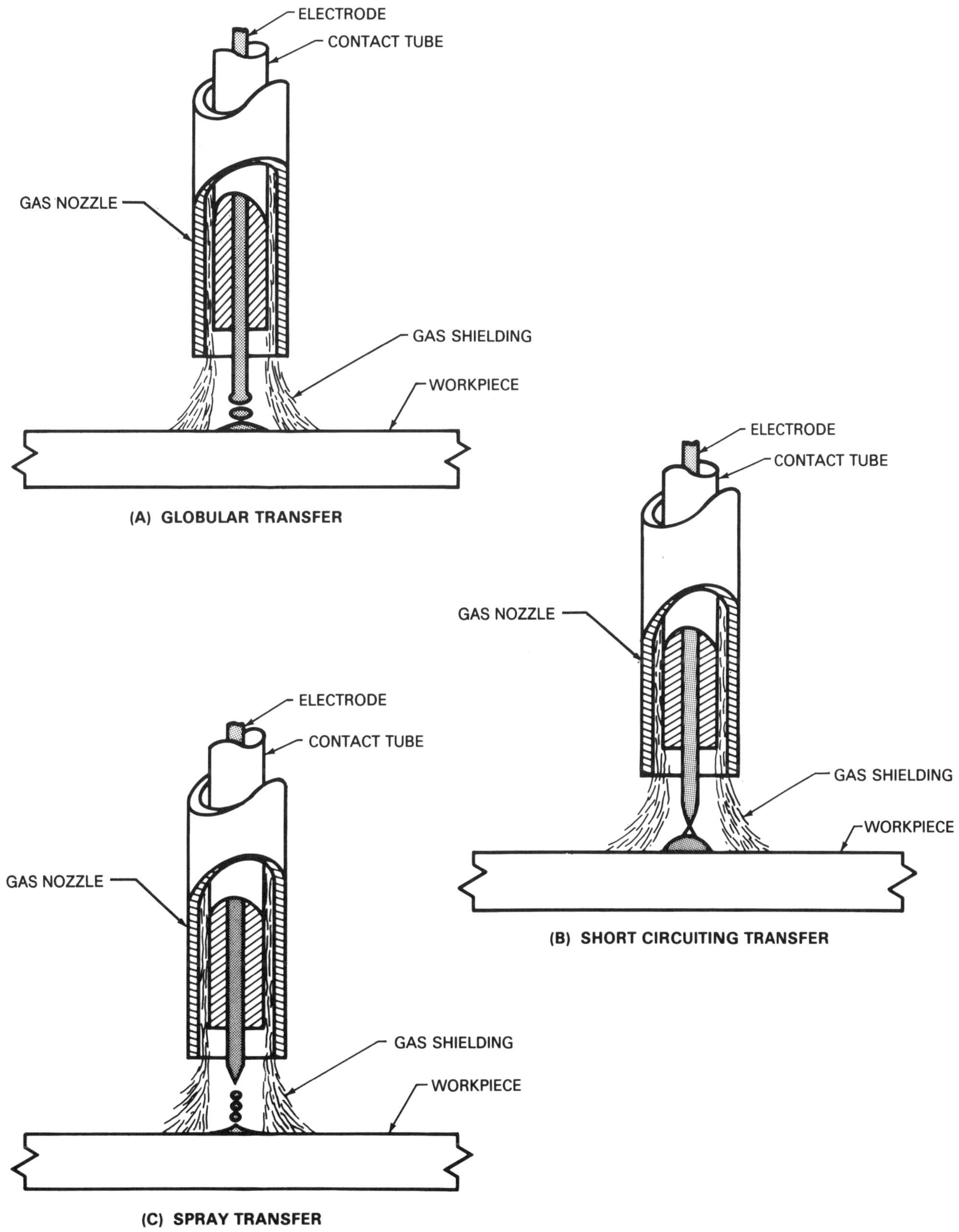

Figure 39 — Metal Transfer in Gas Metal Arc Welding

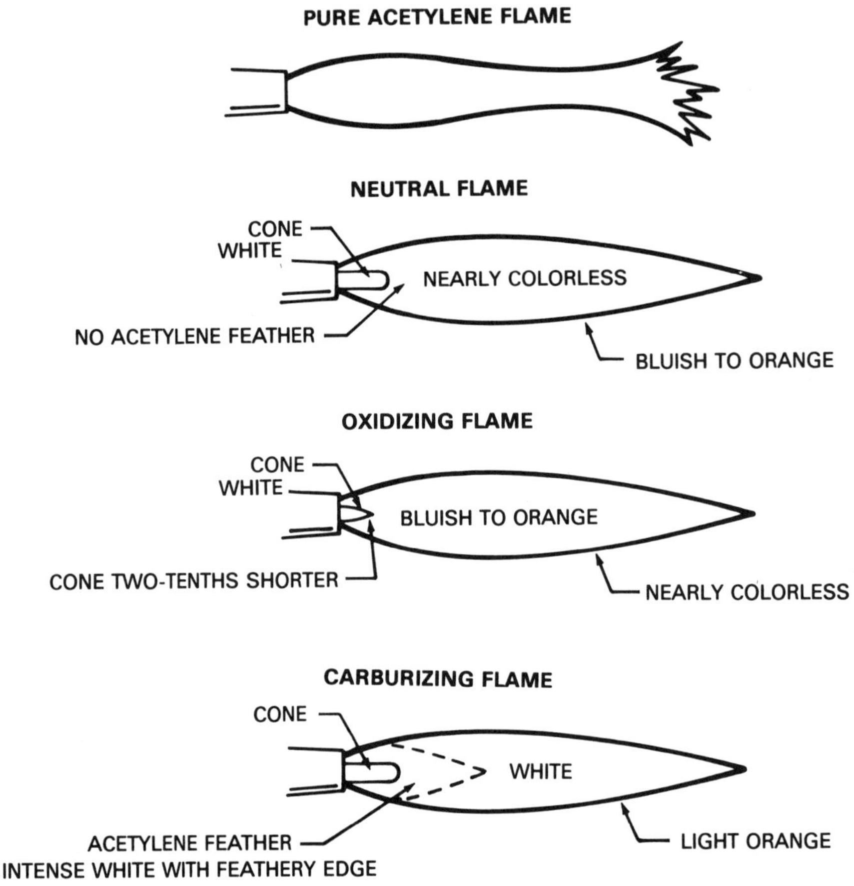

Figure 40 — Oxyacetylene Flame

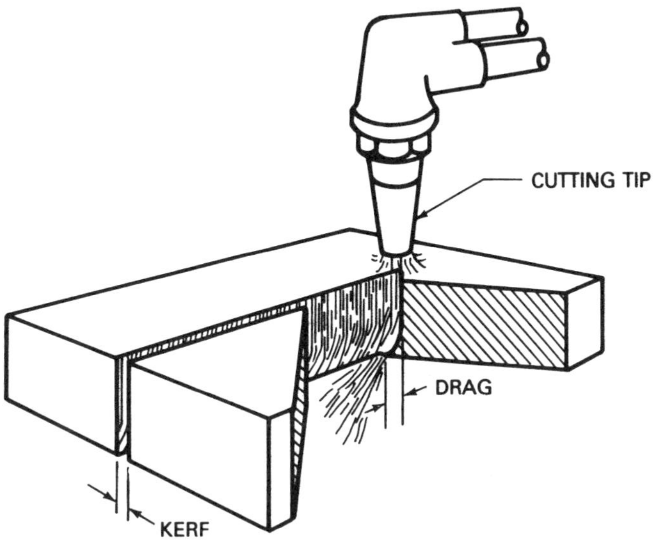

Figure 41 — Oxygen Cutting

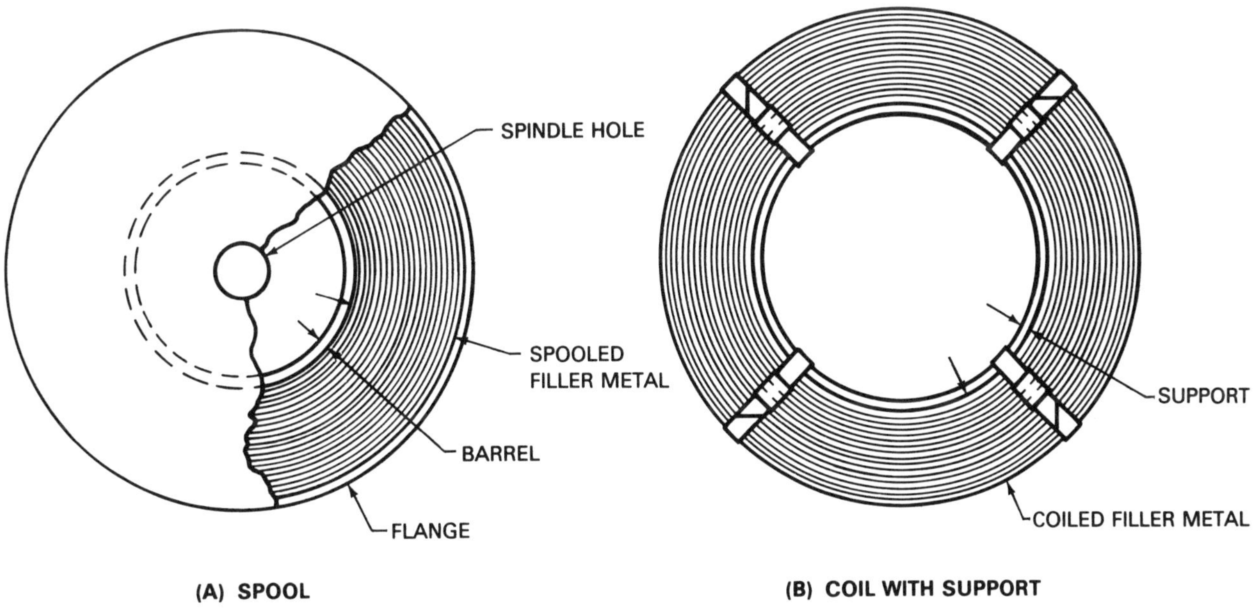

Figure 42 — Filler Metal Packaging

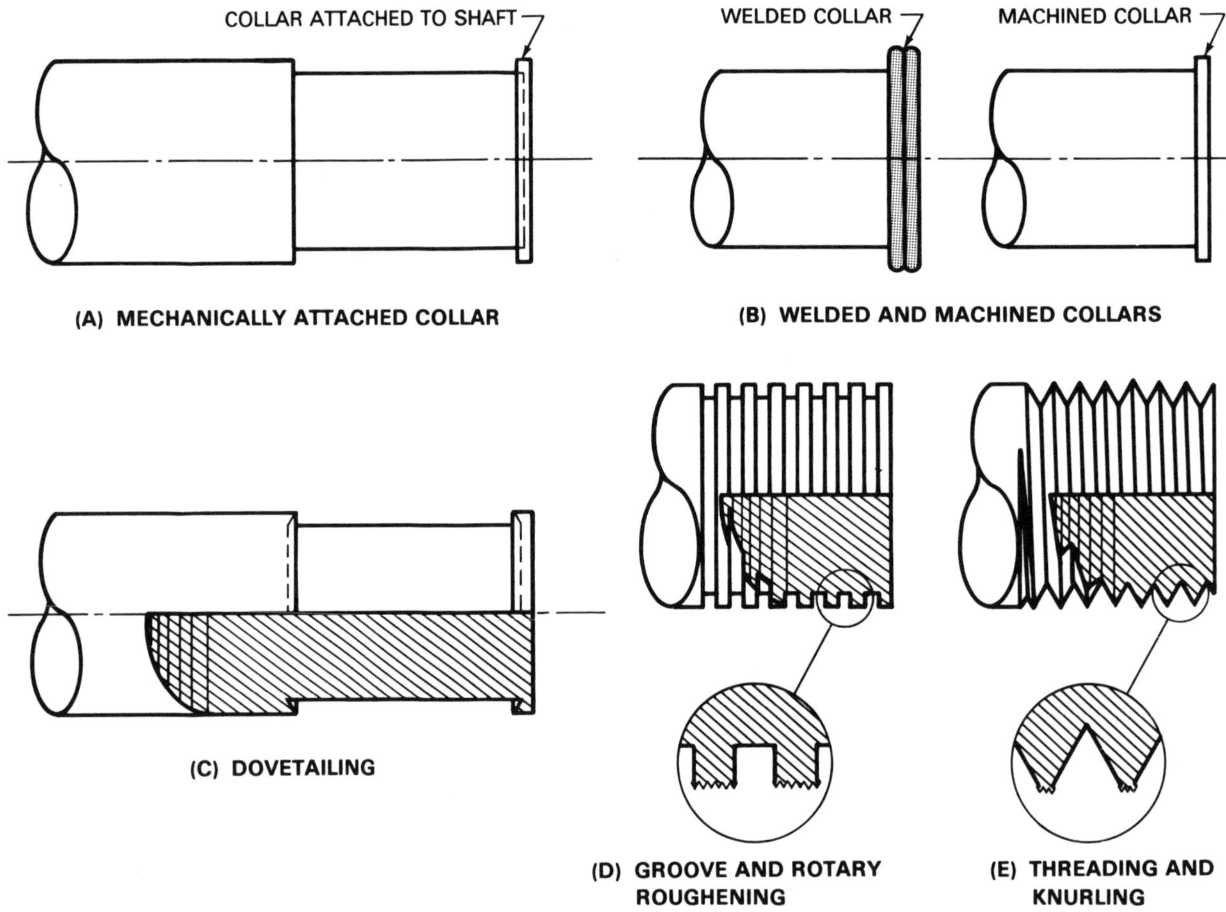

Figure 43 — Thermal Spraying Surface Preparation

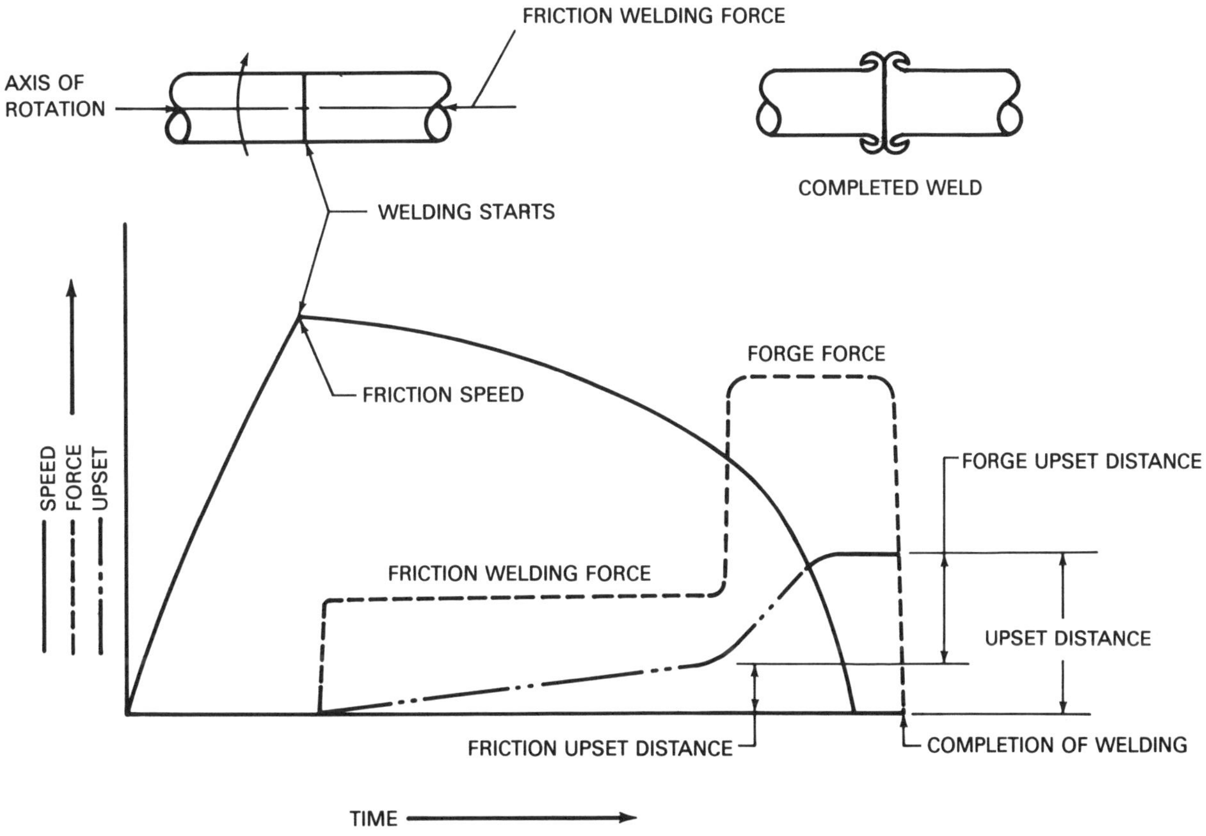

Figure 44 — Generalized Diagram of Inertia Friction Welding

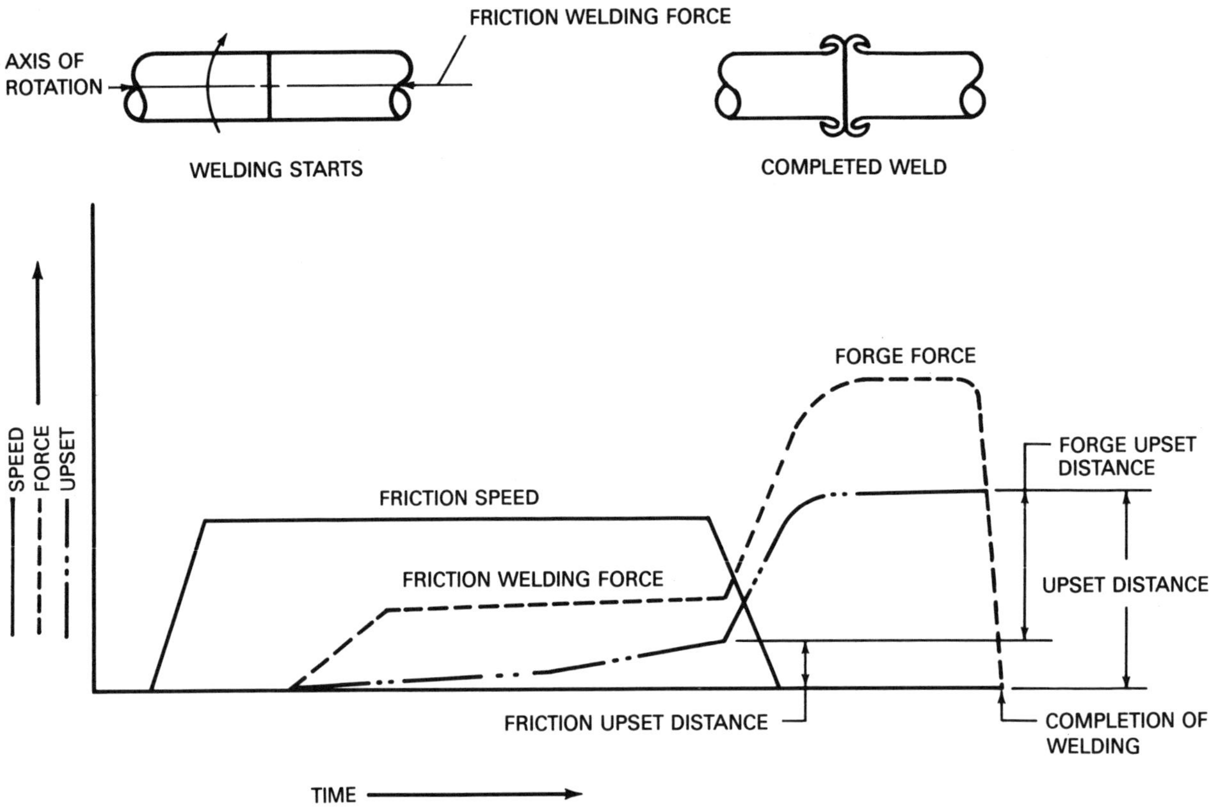

Figure 45 — Generalized Diagram of Direct Drive Friction Welding

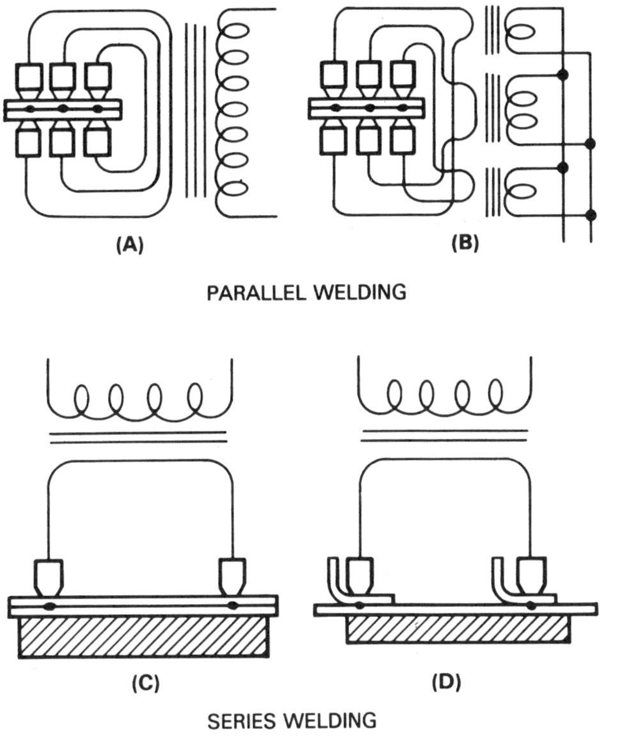

Figure 46 — Typical Arrangements for Multiple Spot Welding

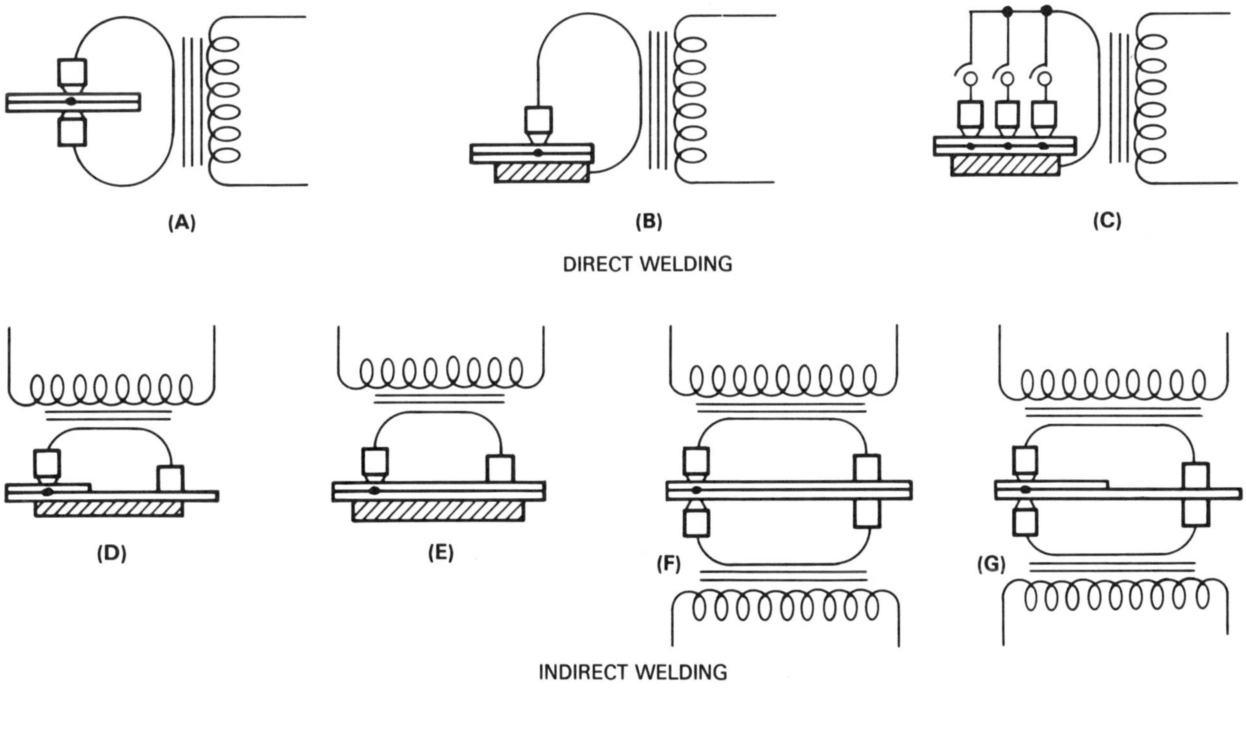

Figure 47 — Typical Arrangements for Single Spot Welds

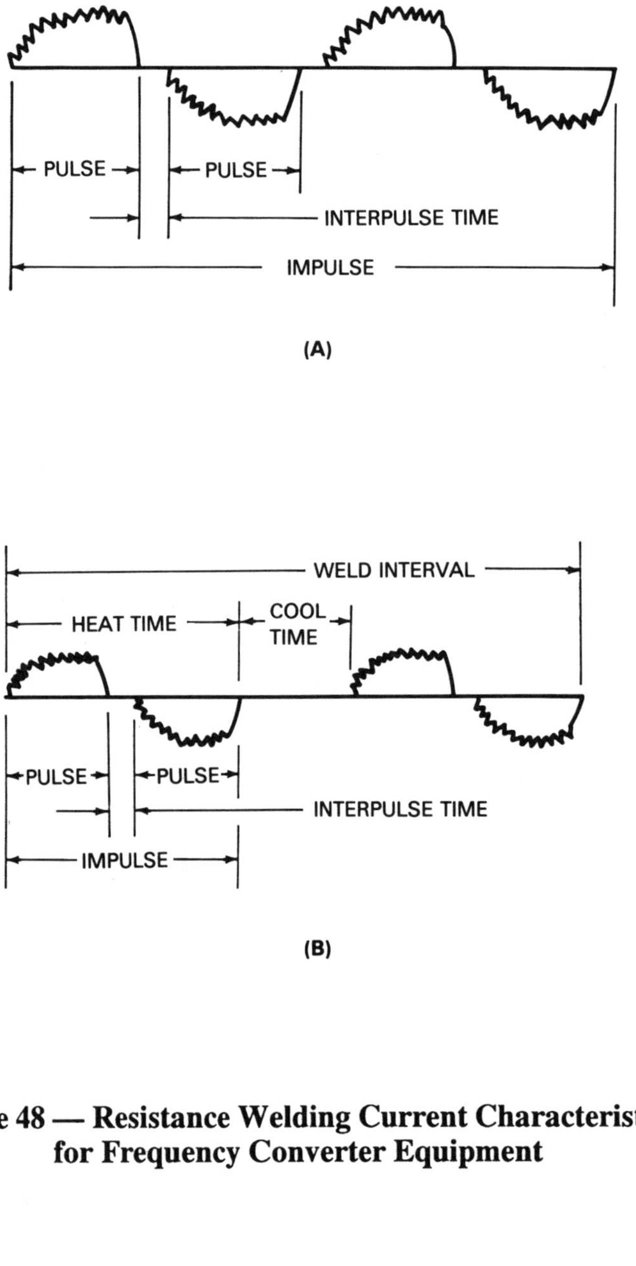

Figure 48 — Resistance Welding Current Characteristics for Frequency Converter Equipment

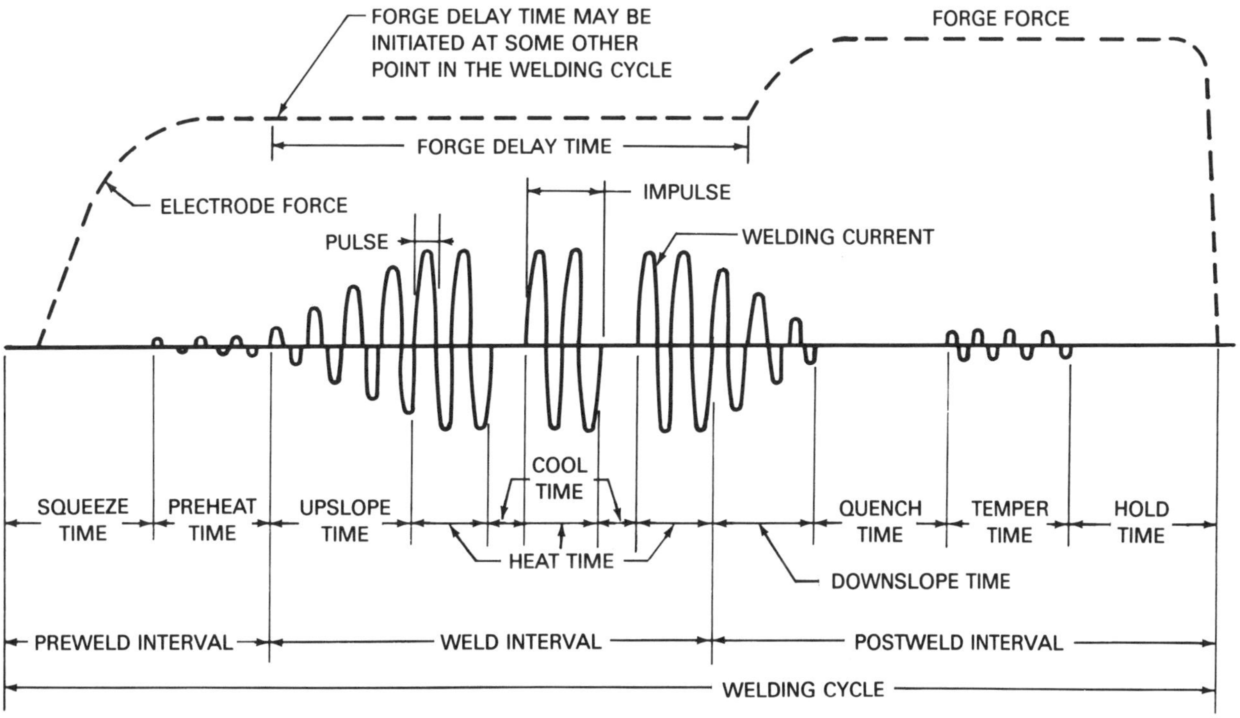

Figure 49 — Multiple-Impulse Resistance Spot Welding Schedule

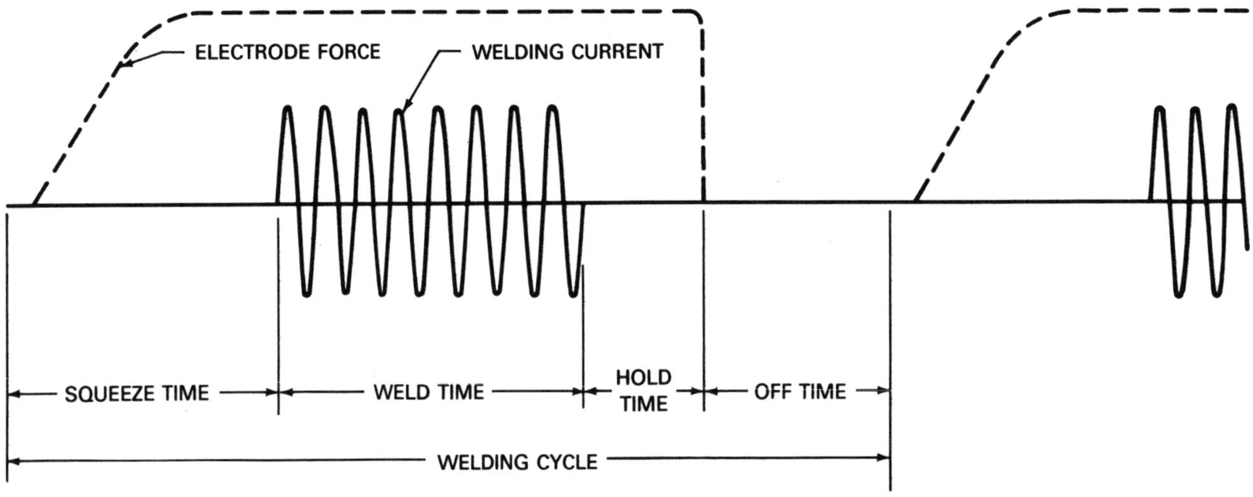

Figure 50 — Single-Impulse Resistance Spot Welding Schedule

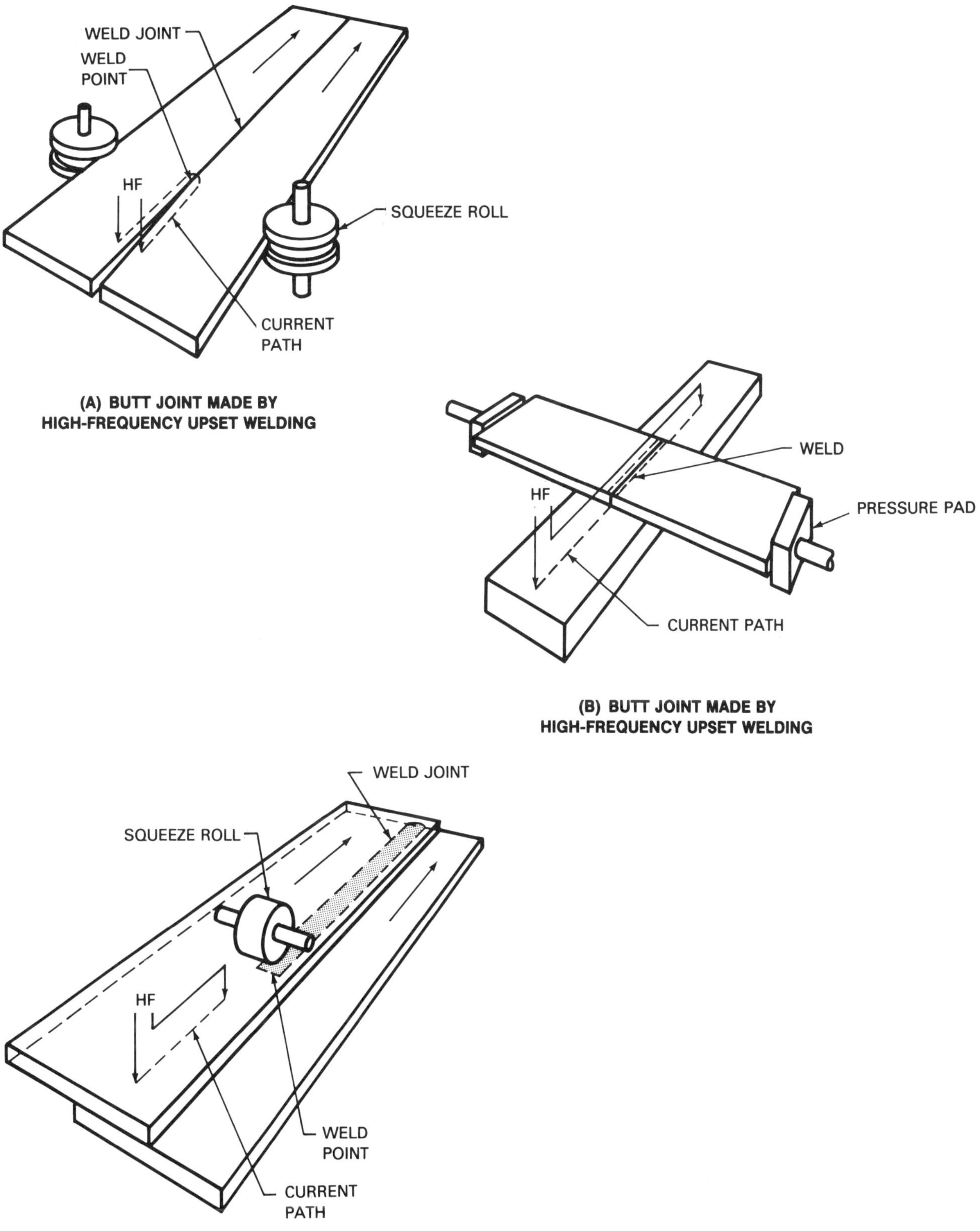

(A) BUTT JOINT MADE BY HIGH-FREQUENCY UPSET WELDING

(B) BUTT JOINT MADE BY HIGH-FREQUENCY UPSET WELDING

(C) LAP JOINT MADE BY HIGH-FREQUENCY SEAM WELDING

Figure 51 — High-Frequency Resistance Welding

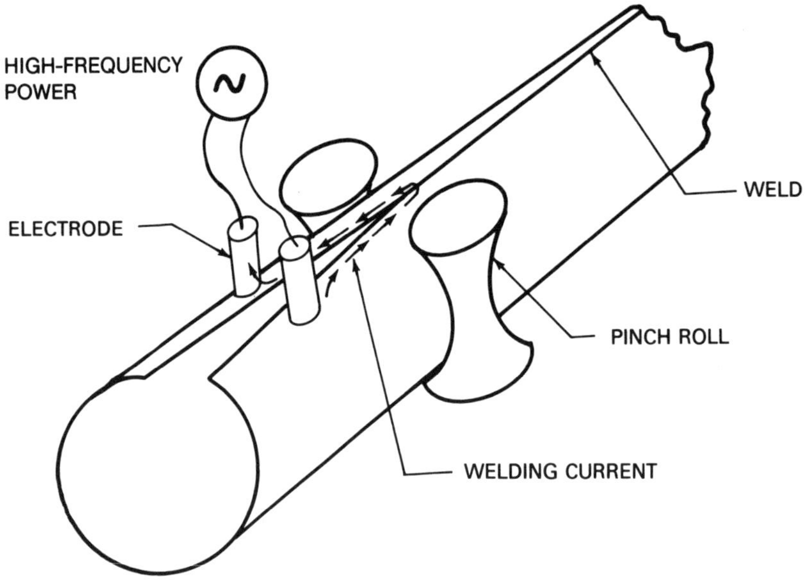

(D) HIGH FREQUENCY UPSET WELDING OF TUBE

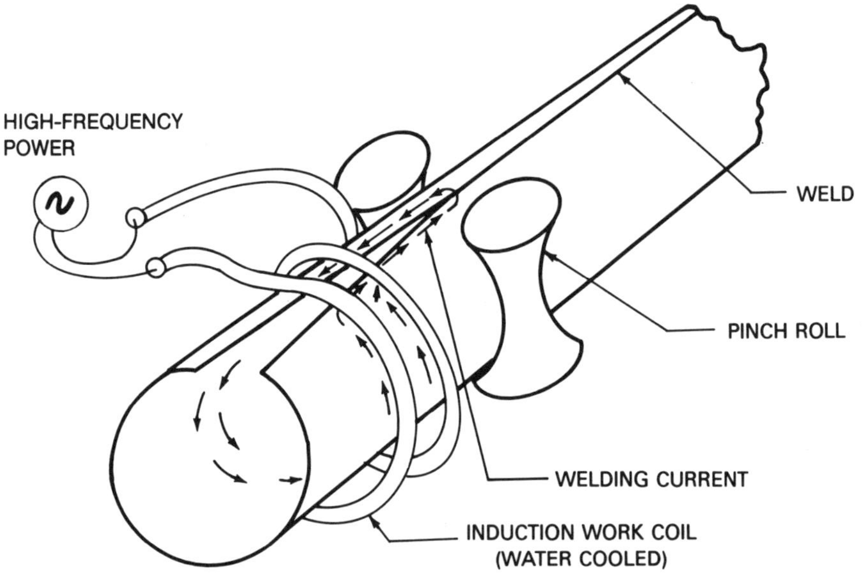

(E) INDUCTION UPSET WELDING OF TUBE

Figure 51 (Continued) — High-Frequency Resistance Welding

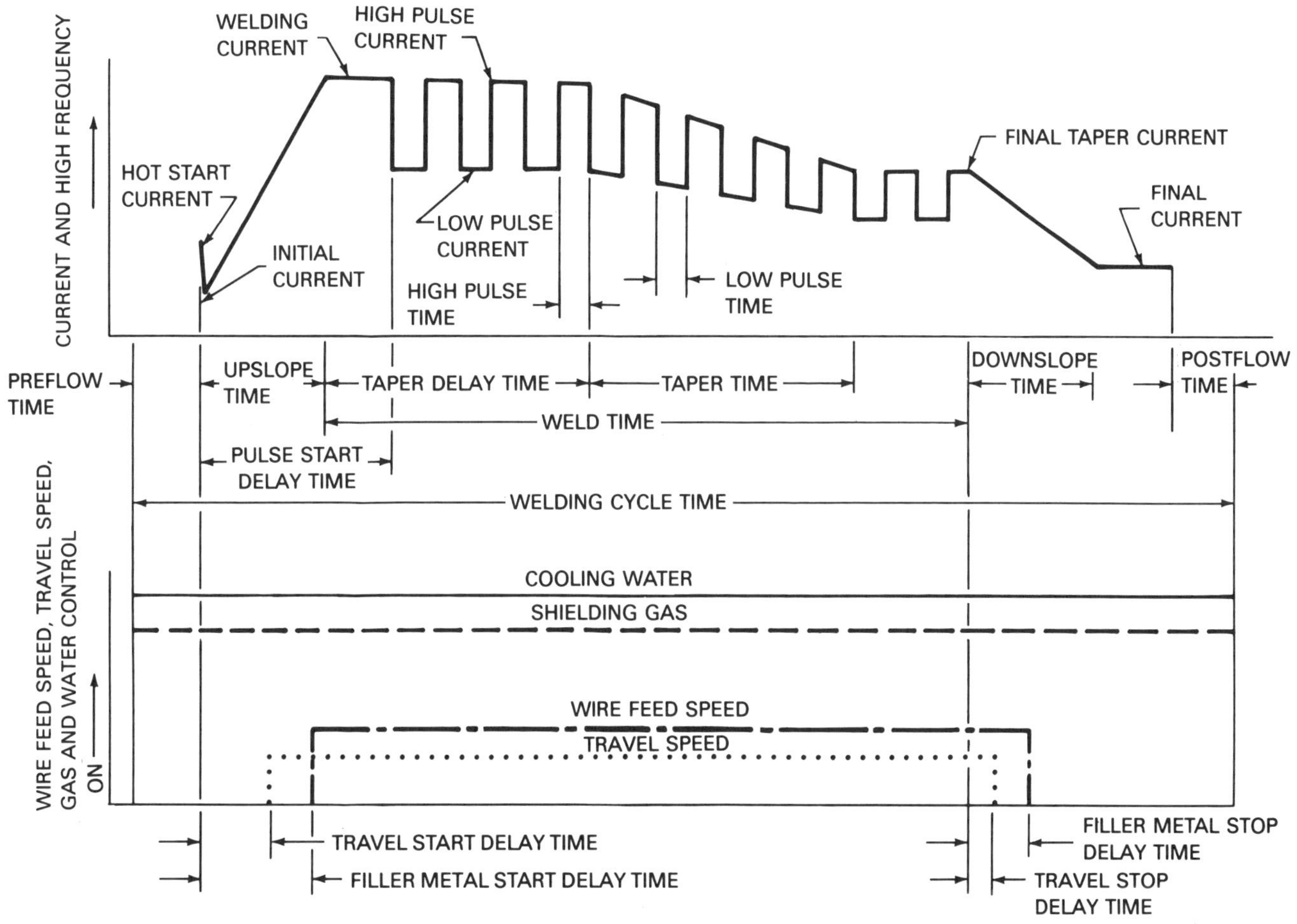

Figure 52 — Typical GTAW or PAW Program for Automatic Welding

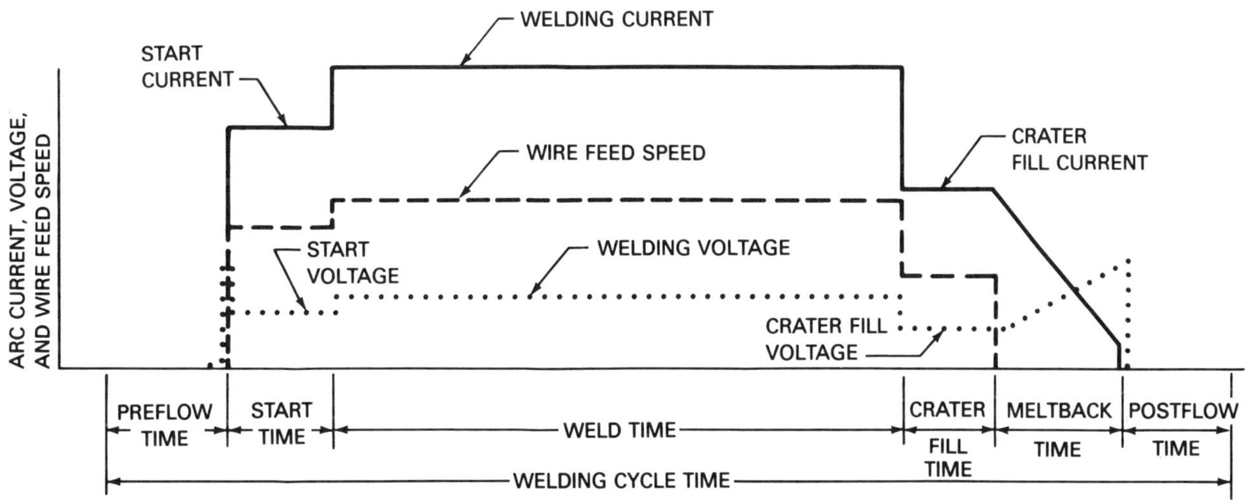

Figure 53 — Typical GMAW, FCAW, and SAW Program for Automatic Welding

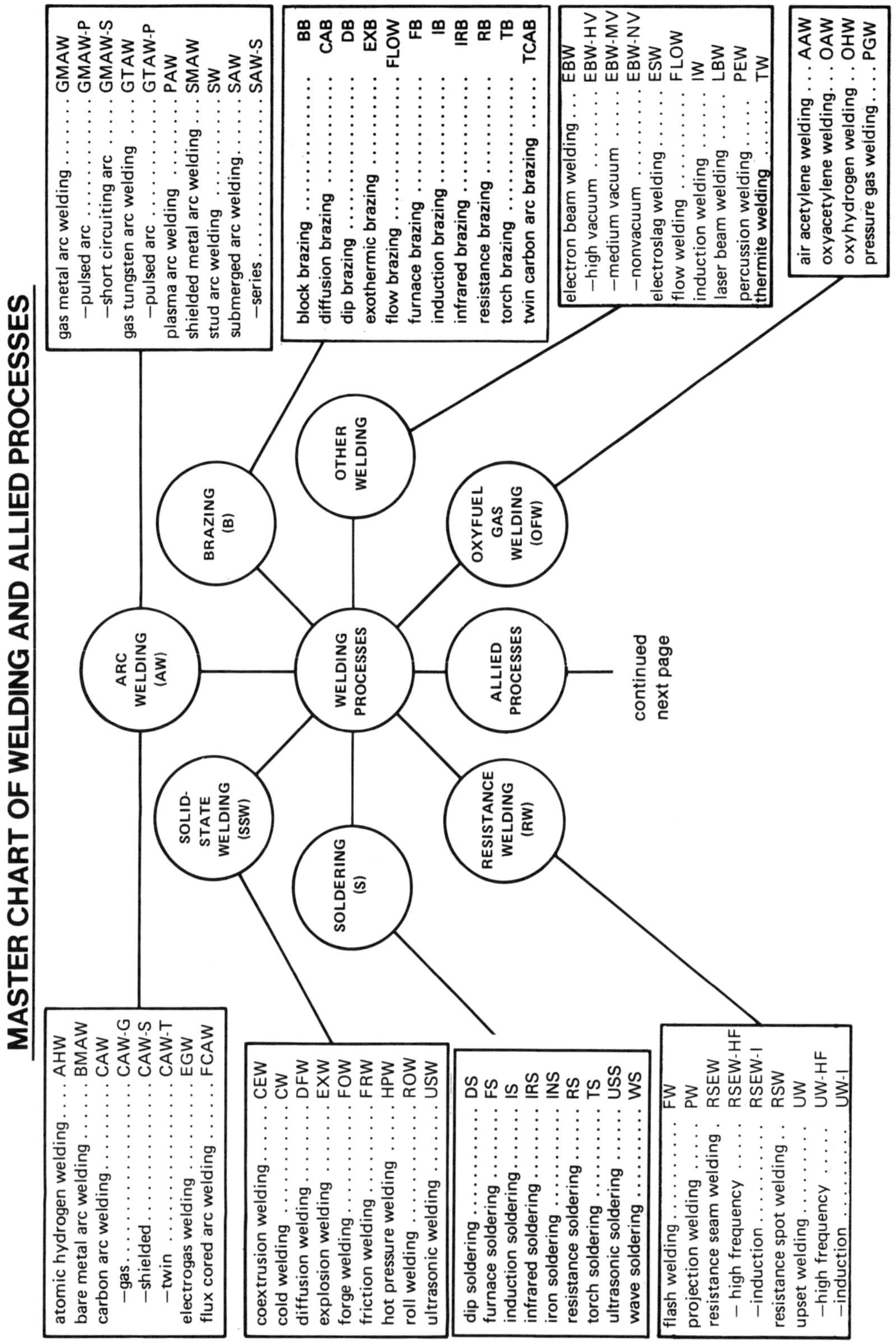

MASTER CHART OF WELDING AND ALLIED PROCESSES

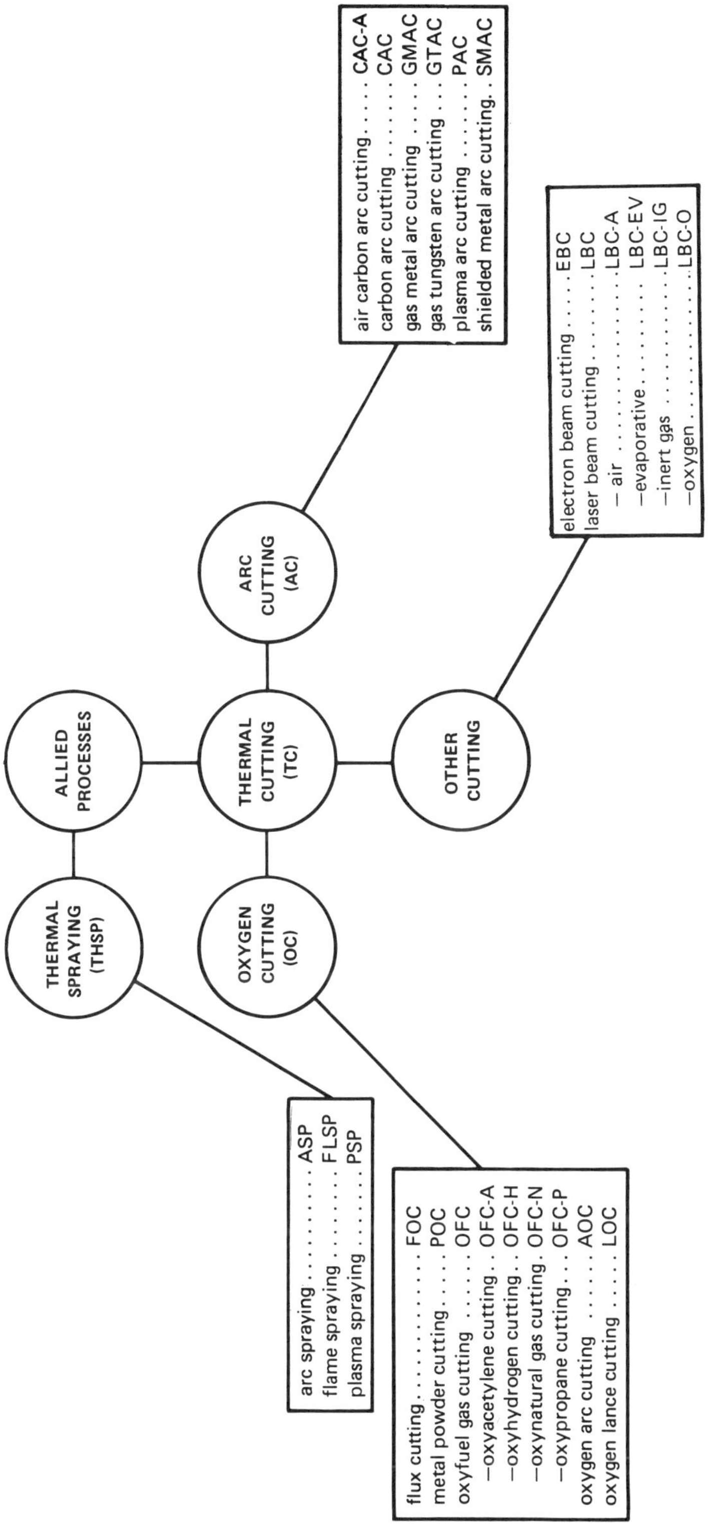

MASTER CHART OF WELDING AND ALLIED PROCESSES (CONT.)

JOINING METHOD CHART

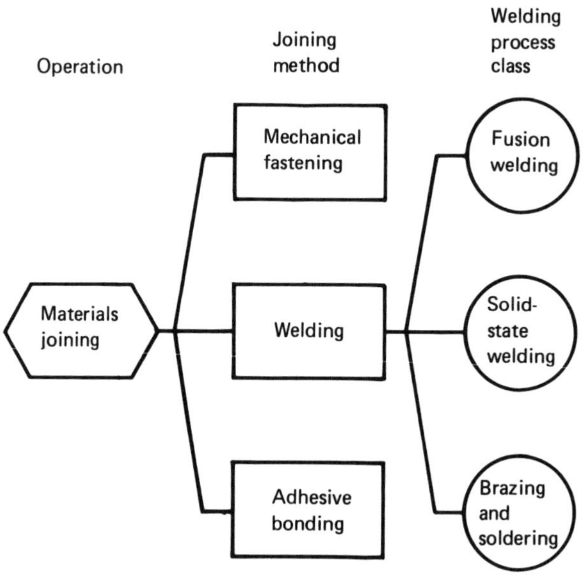

FUSION WELDING CLASSIFICATION CHART

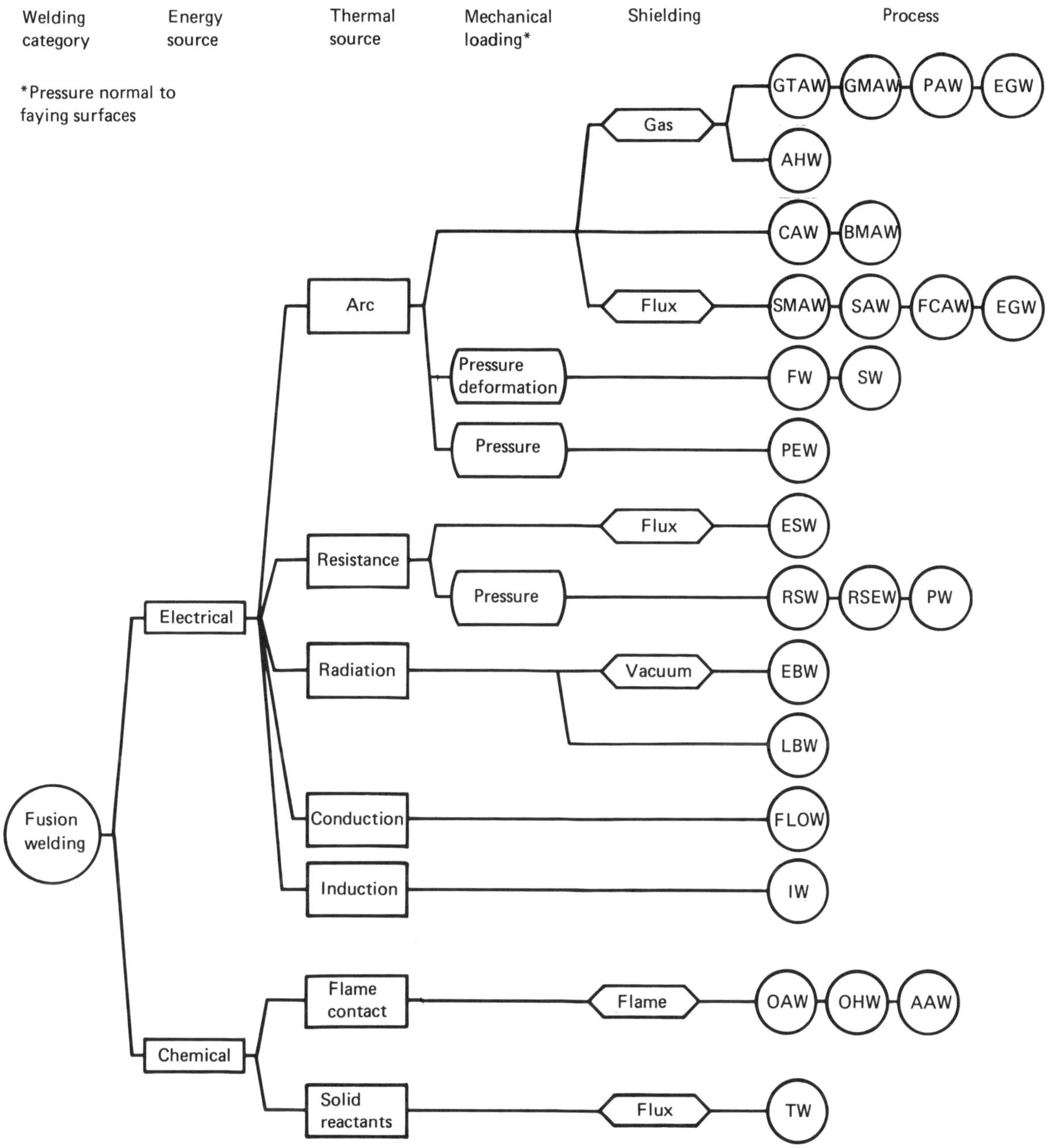

Definitions

Designation	Welding process
AAW	Air acetylene
AHW	Atomic hydrogen
BMAW	Bare metal arc
CAW	Carbon arc
EBW	Electron beam
EGW	Electrogas
ESW	Electroslag
FLOW	Flow
FCAW	Flux cored arc

Definitions

Designation	Welding process
FW	Flash
GMAW	Gas metal arc
GTAW	Gas tungsten arc
IW	Induction
LBW	Laser beam
OAW	Oxyacetylene
OHW	Oxyhydrogen
PAW	Plasma arc
PEW	Percussion

Definitions

Designation	Welding process
PW	Projection
RSEW	Resistance seam
RSW	Resistance spot
SAW	Submerged arc
SMAW	Shielded metal arc
SW	Stud arc
TW	Thermit

SOLID-STATE WELDING CLASSIFICATION CHART

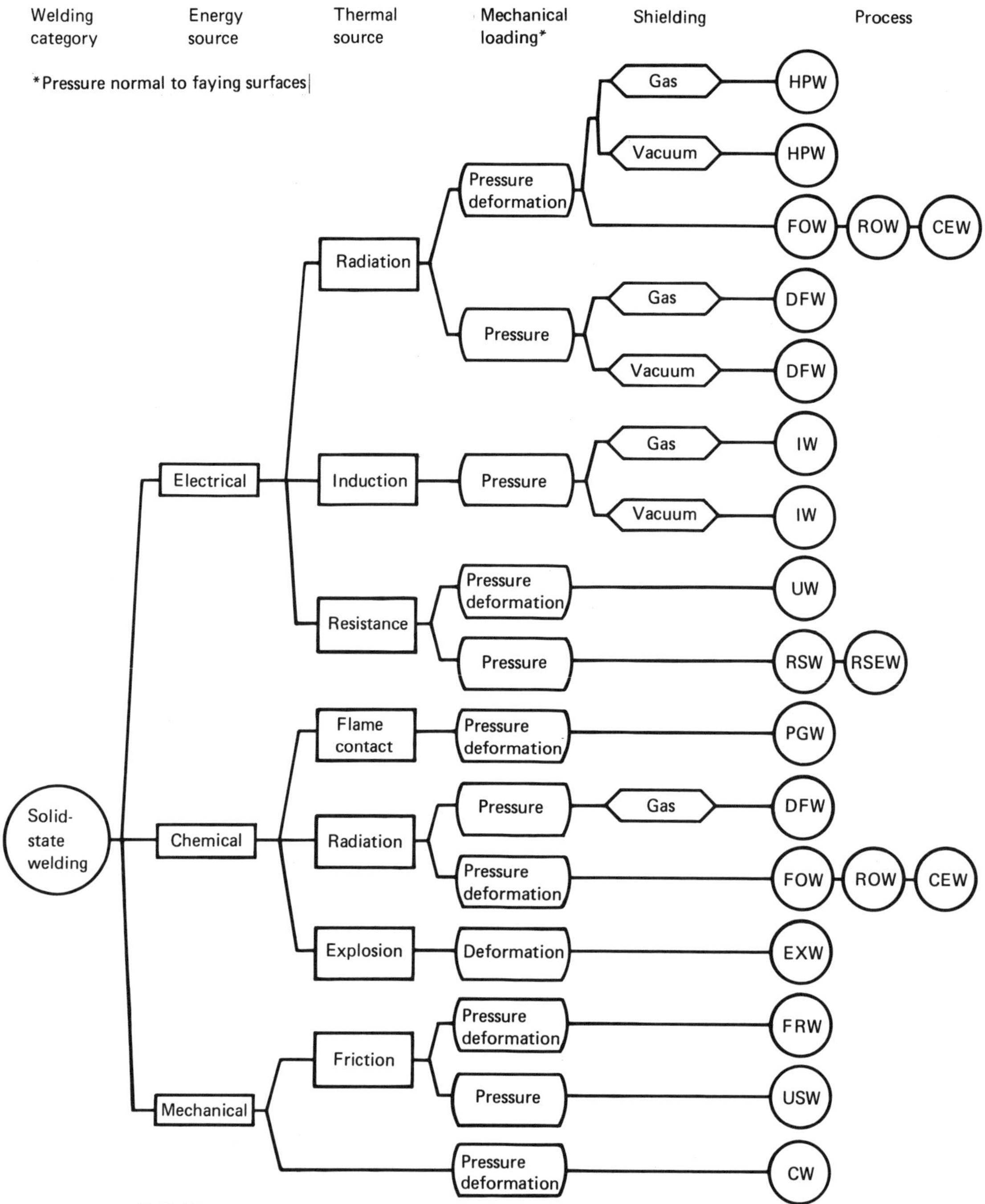

Definitions

Designation	Welding Process
CEW	Coextrusion
CW	Cold
DFW	Diffusion
EXW	Explosion
FOW	Forge
FRW	Friction
HPW	Hot pressure

Definitions

Designation	Welding process
IW	Induction
PGW	Pressure gas
RSEW	Resistance seam
RSW	Resistance spot
ROW	Roll
USW	Ultrasonic
UW	Upset

BRAZING AND SOLDERING CLASSIFICATION CHART

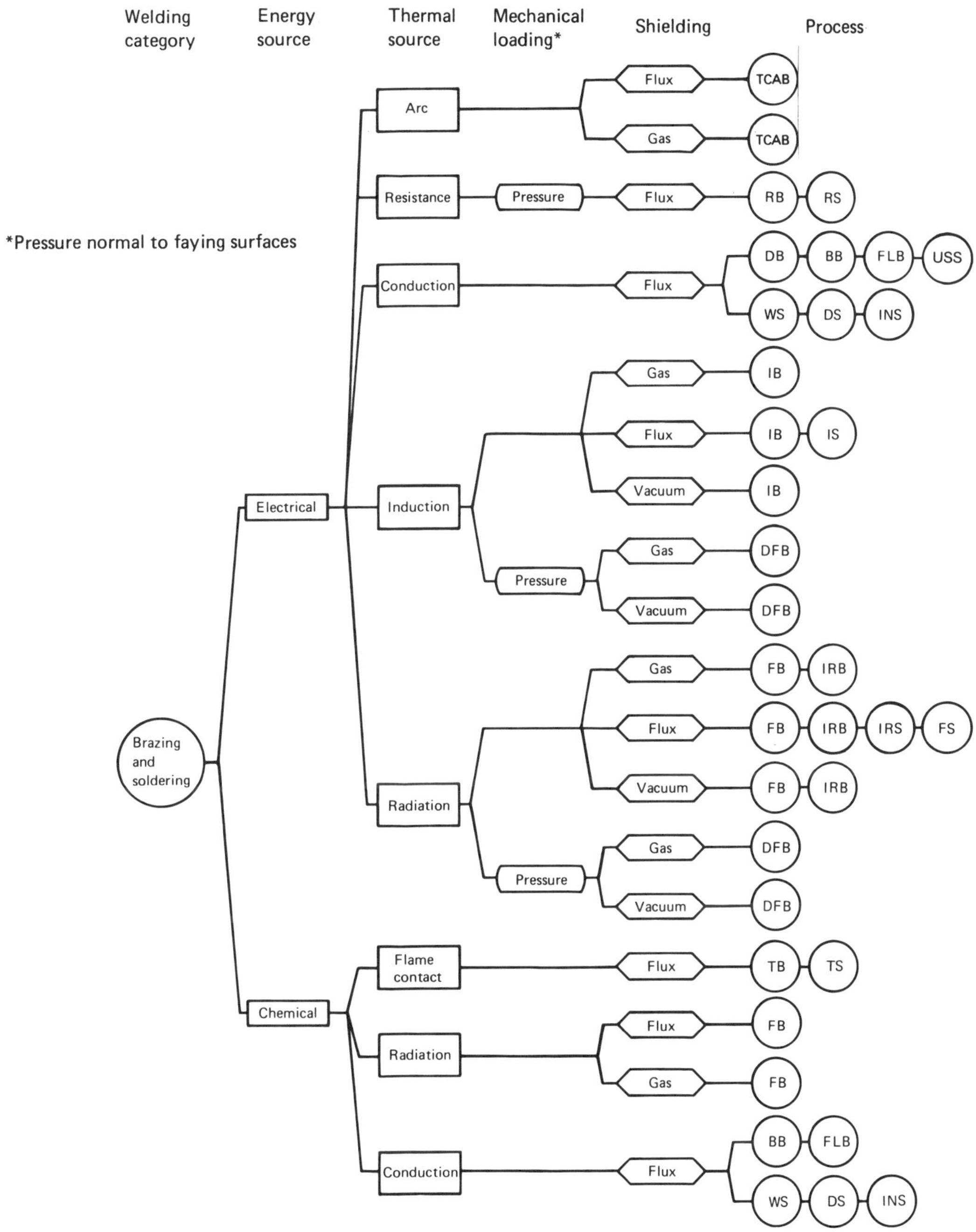

Definitions

Designation	Process
AB	Arc brazing
BB	Block brazing
TCAB	Twin carbon arc brazing
DB	Dip brazing
DS	Dip soldering
DFB	Diffusion brazing
FB	Furnace brazing
FS	Furnace soldering
FLB	Flow brazing
IB	Induction brazing

Definitions

Designation	Process
IS	Induction soldering
IRB	Infrared brazing
IRS	Infrared soldering
INS	Iron soldering
RB	Resistance brazing
RS	Resistance soldering
TB	Torch brazing
TS	Torch soldering
USS	Ultrasonic soldering
WS	Wave soldering

Table 1
Letter Designations of Welding and Allied Processes and their Variations

Processes and Variations	Letter Designations
arc welding	AW
arc stud welding	SW
atomic hydrogen welding	AHW
bare metal arc welding	BMAW
carbon arc welding	CAW
gas carbon arc welding	CAW-G
shielded carbon arc welding	CAW-S
twin carbon arc welding	CAW-T
electrogas welding	EGW
flux cored arc welding	FCAW
gas shielded flux cored arc welding	FCAW-G
self shielded flux cored arc welding	FCAW-S
gas metal arc welding	GMAW
pulsed gas metal arc welding	GMAW-P
short circuit gas metal arc welding	GMAW-S
gas tungsten arc welding	GTAW
pulsed gas tungsten arc welding	GTAW-P
plasma arc welding	PAW
shielded metal arc welding	SMAW
submerged arc welding	SAW
series submerged arc welding	SAW-S
brazing	B
block brazing	BB
diffusion brazing	DFB
dip brazing	DB
exothermic brazing	EXB
flow brazing	FLB
furnace brazing	FB
induction brazing	IB
infrared brazing	IRB
resistance brazing	RB
torch brazing	TB
twin carbon arc brazing	TCAB
braze welding	BW
arc braze welding	ABW
carbon arc braze welding	CABW
exothermic braze welding	EXBW
other welding processes	
electron beam welding	EBW
high vacuum electron beam welding	EBW-HV
medium vacuum electron beam welding	EBW-MV
nonvacuum electron beam welding	EBW-NV
electroslag welding	ESW
flow welding	FLOW
induction welding	IW
laser beam welding	LBW
percussion welding	PEW
thermite welding	TW
oxyfuel gas welding	OFW
air acetylene welding	AAW
oxyacetylene welding	OAW
oxyhydrogen welding	OHW
pressure gas welding	PGW
resistance welding	RW
flash welding	FW
projection welding	PW
resistance seam welding	RSEW
high-frequency seam welding	RSEW-HF
induction seam welding	RSEW-I
resistance spot welding	RSW
upset welding	UW
high-frequency upset welding	UW-HF
induction upset welding	UW-I
soldering	S
dip soldering	DS
furnace soldering	FS
induction soldering	IS
infrared soldering	IRS
iron soldering	INS
resistance soldering	RS
torch soldering	TS
ultrasonic soldering	USS
wave soldering	WS
solid-state welding	SSW
coextrusion welding	CEW
cold welding	CW
diffusion welding	DFW
explosion welding	EXW
forge welding	FOW
friction welding	FRW
hot pressure welding	HPW
roll welding	ROW
ultrasonic welding	USW
thermal cutting	TC
arc cutting	AC
air carbon arc cutting	CAC-A
carbon arc cutting	CAC
gas metal arc cutting	GMAC
gas tungsten arc cutting	GTAC
plasma arc cutting	PAC
shielded metal arc cutting	SMAC
electron beam cutting	EBC
laser beam cutting	LBC
laser beam air cutting	LBC-A
laser beam evaporative cutting	LBC-EV
laser beam inert gas cutting	LBC-IG
laser beam oxygen cutting	LBC-O
oxygen cutting	OC
flux cutting	FOC
metal powder cutting	POC
oxyfuel gas cutting	OFC
oxyacetylene cutting	OFC-A
oxyhydrogen gas cutting	OFC-H
oxynatural gas cutting	OFC-N
oxypropane cutting	OFC-P
oxygen arc cutting	AOC
oxygen lance cutting	LOC
thermal spraying	THSP
arc spraying	ASP
flame spraying	FLSP
plasma spraying	PSP

Table 2
Alphabetical Cross Reference to Table 1 by Process

Processes and Variations	Letter Designations
arc braze welding	ABW
arc cutting	AC
arc spraying	ASP
arc stud welding	SW
arc welding	AW
air acetylene welding	AAW
air carbon arc cutting	CAC-A
atomic hydrogen welding	AHW
bare metal arc welding	BMAW
block brazing	BB
brazing	B
braze welding	BW
carbon arc braze welding	CABW
carbon arc cutting	CAC
carbon arc welding	CAW
coextrusion welding	CEW
cold welding	CW
diffusion brazing	DFB
diffusion welding	DFW
dip brazing	DB
dip soldering	DS
electrogas welding	EGW
electron beam cutting	EBC
electron beam welding	EBW
electroslag welding	ESW
exothermic brazing	EXB
exothermic braze welding	EXBW
explosion welding	EXW
flame spraying	FLSP
flash welding	FW
flow brazing	FLB
flow welding	FLOW
flux cored arc welding	FCAW
flux cutting	FOC
forge welding	FOW
friction welding	FRW
furnace brazing	FB
furnace soldering	FS
gas carbon arc welding	CAW-G
gas metal arc cutting	GMAC
gas metal arc welding	GMAW
gas shielded flux cored arc welding	FCAW-G
gas tungsten arc cutting	GTAC
gas tungsten arc welding	GTAW
high-frequency seam welding	RSEW-HF
high-frequency upset welding	UW-HF
high vacuum electron beam welding	EBW-HV
hot pressure welding	HPW
induction brazing	IB
induction seam welding	RSEW-I
induction soldering	IS
induction upset welding	UW-I
induction welding	IW
infrared brazing	IRB
infrared soldering	IRS

Processes and Variations	Letter Designations
iron soldering	INS
laser beam air cutting	LBC-A
laser beam cutting	LBC
laser beam evaporative cutting	LBC-EV
laser beam inert gas cutting	LBC-IG
laser beam oxygen cutting	LBC-O
laser beam welding	LBW
medium vacuum electron beam welding	EBW-MV
metal powder cutting	POC
nonvacuum electron beam welding	EBW-NV
oxyacetylene cutting	OFC-A
oxyacetylene welding	OAW
oxyfuel gas cutting	OFC
oxyfuel gas welding	OFW
oxygen arc cutting	AOC
oxygen cutting	OC
oxygen lance cutting	LOC
oxyhydrogen cutting	OFC-H
oxyhydrogen welding	OHW
oxynatural gas cutting	OFC-N
oxypropane cutting	OFC-P
percussion welding	PEW
plasma arc cutting	PAC
plasma arc welding	PAW
plasma spraying	PSP
pressure gas welding	PGW
projection welding	PW
pulsed gas metal arc welding	GMAW-P
pulsed gas tungsten arc welding	GTAW-P
resistance brazing	RB
resistance seam welding	RSEW
resistance soldering	RS
resistance spot welding	RSW
resistance welding	RW
roll welding	ROW
self shielded flux cored arc welding	FCAW-S
series submerged arc welding	SAW-S
shielded carbon arc welding	CAW-S
shielded metal arc cutting	SMAC
shielded metal arc welding	SMAW
short circuit gas metal arc welding	GMAW-S
soldering	S
solid-state welding	SSW
submerged arc welding	SAW
thermal cutting	TC
thermal spraying	THSP
thermite welding	TW
torch brazing	TB
torch soldering	TS
twin carbon arc brazing	TCAB
twin carbon arc welding	CAW-T
ultrasonic welding	USW
ultrasonic soldering	USS
upset welding	UW
wave soldering	WS

Table 3
Alphabetical Cross Reference to Table 1 by Letter Designation

Processes and Variations	Letter Designations
AAW	air acetylene welding
ABW	arc braze welding
AC	arc cutting
AHW	atomic hydrogen welding
AOC	oxygen arc cutting
ASP	arc spraying
AW	arc welding
B	brazing
BB	block brazing
BMAW	bare metal arc welding
BW	braze welding
CABW	carbon arc braze welding
CAC	carbon arc cutting
CAC-A	air carbon arc cutting
CAW	carbon arc welding
CAW-G	gas carbon arc welding
CAW-S	shielded carbon arc welding
CAW-T	twin carbon arc welding
CEW	coextrusion welding
CW	cold welding
DB	dip brazing
DFB	diffusion brazing
DFW	diffusion welding
DS	dip soldering
EBC	electron beam cutting
EBW	electron beam welding
EBW-HV	high vacuum electron beam welding
EBW-MV	medium vacuum electron beam welding
EBW-NV	nonvacuum electron beam welding
EGW	electrogas welding
ESW	electroslag welding
EXB	exothermic brazing
EXBW	exothermic braze welding
EXW	explosion welding
FB	furnace brazing
FCAW	flux cored arc welding
FCAW-G	gas shielded flux cored arc welding
FCAW-S	self-shielded flux cored arc welding
FLB	flow brazing
FLOW	flow welding
FLSP	flame spraying
FOC	flux cutting
FOW	forge welding
FRW	friction welding
FS	furnace soldering
FW	flash welding
GMAC	gas metal arc cutting
GMAW	gas metal arc welding
GMAW-P	pulsed gas metal arc welding
GMAW-S	short circuit gas metal arc welding
GTAC	gas tungsten arc cutting
GTAW	gas tungsten arc welding
GTAW-P	pulsed gas tungsten arc welding
HPW	hot pressure welding
IB	induction brazing
INS	iron soldering
IRB	infrared brazing
IRS	infrared soldering
IS	induction soldering
IW	induction welding
LBC	laser beam cutting
LBC-A	laser beam air cutting
LBC-EV	laser beam evaporative cutting
LBC-IG	laser beam inert gas cutting
LBC-O	laser beam oxygen cutting
LBW	laser beam welding
LOC	oxygen lance cutting
OAW	oxyacetylene welding
OC	oxygen cutting
OFC	oxyfuel gas cutting
OFC-A	oxyacetylene cutting
OFC-H	oxyhydrogen cutting
OFC-N	oxynatural gas cutting
OFC-P	oxypropane cutting
OFW	oxyfuel gas welding
OHW	oxyhydrogen welding
PAC	plasma arc cutting
PAW	plasma arc welding
PEW	percussion welding
PGW	pressure gas welding
POC	metal powder cutting
PSP	plasma spraying
PW	projection welding
RB	resistance brazing
ROW	roll welding
RS	resistance soldering
RSEW	seam welding
RSEW-HF	high-frequency seam welding
RSEW-I	induction seam welding
RSW	resistance spot welding
RW	resistance welding
S	soldering
SAW	submerged arc welding
SAW-S	series submerged arc welding
SMAC	shielded metal arc cutting
SMAW	shielded metal arc welding
SSW	solid-state welding
SW	arc welding stud
TB	torch brazing
TC	thermal cutting
TCAB	twin carbon arc brazing
THSP	thermal spraying
TS	torch soldering
TW	thermite welding
USS	ultrasonic soldering
USW	ultrasonic welding
UW	upset welding
UW-HF	high-frequency upset welding
UW-I	induction upset welding
WS	wave soldering

Table 4
Suffixes for Optional Use in Applying Welding and Allied Processes

Adaptive control	AD	Mechanized	ME
Automatic	AU	Robotic	RO
Manual	MA	Semiautomatic	SA

Table 5
Obsolete or Seldom Used Processes

Welding Process or Variation	Letter Designations	Welding Process or Variation	Letter Designations
Air acetylene welding	AAW	Flow brazing	FLB
Atomic hydrogen welding	AHW	Flow welding	FLOW
Bare metal arc welding	BMAW	Twin carbon arc brazing	TCAB
Block brazing	BB	Gas carbon arc welding	CA

Annex A

The Definitions Subcommittee Manifesto of 1987

(This Annex is not a part of ANSI/AWS A3.0-94, *Standard Welding Terms and Definitions,* but is included for informational purposes only.)

Note: *Webster's Ninth New Collegiate Dictionary,* 1986, defines *manifesto* as, "a public declaration of intentions, motives, or views." It is in this sense that the term is used.

The primary purpose of language is the communication of concepts. Because welding deals with sometimes complicated and not everyday concepts, it is essential to use language that can express those concepts clearly, accurately, and unambiguously. Unambiguity requires a common terminology throughout the welding industry. It is also essential that each term have only one meaning, so that a reader does not have to guess the meaning. To reduce the specialized vocabulary one must learn to understand welding information, it is desirable that each concept represent one term only. A common welding terminology should certainly be a practical goal, when one considers how insignificant the entire welding world is, in comparison to all the areas of human endeavor.

A welding terminology that attempts to conform to all of the characteristics of good terminology is the responsibility of the American Welding Society, as represented by their Definitions Subcommittee of the Committee on Definitions and Symbols. That welding terminology is published as ANSI/AWS A3.0, *Standard Welding Terms and Definitions,* written to promote consistent and accurate usage of welding terms in standards, technical papers, and other written or spoken information pertaining to welding.

Good welding terminology is many things. Consistency in the use and meaning of a term in the welding industry is one of the desirable characteristics. However, the rather uncontrolled evolution and growth of welding terms in the past makes the application of that principle very difficult, and sometimes impossible. The Definitions Subcommittee sometimes finds two or more popular terms for the same thing or, even worse, two meanings for the same term. Some popular terms are simply incorrect, and not in some nitpicking detail, but in violation of fundamental dictates of logic or grammar. The majority is usually right; but not always.

Good welding terminology also requires the elimination of unnecessary words in definitions. The fewer words, the better the understanding of those words — within limits of course. Verbiage is avoided by the practice of a number of principles. Terms consisting of word combinations, where the definitions of their elements (found in either the dictionary or in A3.0) make the meaning of the combination clear, should not be included. Definitions should be written to include as many uses as possible, while still retaining clarity and accuracy, but should not be extended to every nuance of meaning. A uniform format is used for certain groups of terms.

One need not belabor the point that logic, technical accuracy, correct grammar, and simplicity are essential to good welding terminology.

To prevent welding terminology from becoming cumbersome or voluminous, terms should be directly related to welding or allied fields. Terms that are adequately defined in the English dictionary should not be included. Adequate definition requires that there be only one clearly applicable definition, and that definition must accurately reflect the term's use in the welding world. For each definition, there should be only one term, and for each term, there must be only one definition, except where they are distinguished by a delimiting expression.

Definitions are not intended to replace portions of textbooks or specifications, but rather, are intended to ensure that the meaning of each term used in those documents is clear and is the same for all readers. Further information on the guiding principles used by the Definitions Subcommittee is given in Annex B. Knowledge of those principles by the users of welding terminology will increase their understanding of terms and definitions in A3.0, improve the effectiveness of their suggestions and criticisms, and make the Definitions Subcommittee more accountable for their terminology decisions.

The consideration of most welding terms and their definitions involves opinions from many sources. In addition to those of the Definitions Subcommittee members, advice is sought and received from other Committees of the American Welding Society and from organizations outside the AWS. In the resolution of those diverse opinions, the Definitions Subcommittee attempts to maintain in the resulting terms and definitions, all of the desirable attributes of good welding terminology. That ideal is sometimes not attainable because of various factors, such as entrenched use, disagreement among our own members, pressure from those with parochial interests, or human fallibility. The result is often a compromise between the sometimes incompatible characteristics of good welding terminology. Where there is no clear superiority between competing versions of a given term or its definition, the Definitions Subcommittee has no choice but to make a somewhat arbitrary decision.

Of perhaps more importance than absolute logic, precise conformance to the English language, long established usage, or personal preference is that a given term have the same meaning regardless of where, when, or by whom it is used. That can be assured only by a welding terminology common to all. An orderly and efficient transfer of information cannot be expected if the welder uses a different language than the engineer, or if the bridge builder does not understand the pipeline specialist. Who is in a better position to evaluate, introduce, select, modify, solve conflicts, or otherwise do the work necessary to produce those welding terms and definitions for the entire welding fraternity than the group, the Definitions Subcommittee, established by the American Welding Society for that sole purpose?

In recognition of the fact that welding terminology is not static, the Definitions Subcommittee has under constant study and review many new terms and modifications of current terms, applying to the entire spectrum of the welding industry. Those terms and changes that will improve welding terminology and its understanding are introduced into the welding vocabulary. However, those changes must be made only after careful consideration. In addition to the appraisal of a proposed new term or change on its own merits, one must evaluate its effect on interrelated terms, to maintain coherence, and to avoid duplication and conflicts. That may require changes in several terms. Some of those changes may be deemed unacceptable, forcing an adjustment or rejection of the proposed new term or change. The desirability of a proposed change must be balanced against the advantages of a stable terminology. Change can sometimes cause more confusion than that caused by some minor flaw, either real or fancied, in the current term or definition.

It is one of the goals of the Definitions Subcommittee that A3.0 encompass all terms, not adequately defined in the dictionary, directly related to welding or allied fields. Nothing is excluded, except scatology. Both standard and nonstandard jargon, as well as dialect and vernacular terms, are accepted for inclusion in A3.0. Many nonstandard terms already may be found in A3.0, and the remainder will be added. The inclusion of nonstandard terms serves a dual purpose: informational and educational. Since A3.0 is intended as a comprehensive source of welding terms, it must include all of the terms found in welding prose. The definition of most nonstandard terms reveals the corresponding standard term.

The Definitions Subcommittee has been criticized for a lack of accommodation of the welder's language. A few of the most notable deviations are chosen in support of that criticism, but when the entire body of standard terms is considered, most of those terms are the same as those used by the welder. We must make exceptions however, when the vernacular term is clearly incorrect or conflicts with other welding terms. We reject the premise that the welder is incapable of learning those exceptions.

Let us now examine in detail some of the terminology decisions that have long been under attack. Three of the most controversial are welder versus weldor, gas tungsten arc welding (or gas metal arc welding) versus TIG (or MIG), and diffusion welding versus diffusion bonding.

The use of the term *welder* to indicate the person who does the welding originated in the early days of welding and has been reaffirmed by the American Welding Society for the past thirty years. To distinguish the welder from the machine used to perform the welding, the term *welding machine* was introduced for the latter. On the other hand, it has been claimed by some that welding terminology would be improved by substitution of the term "welder" for the term "welding machine" and the term "weldor" for the term "welder". That has the advantage of greater simplicity (but not much); and the written terms are clearly distinguishable. However, that ignores the spoken language. While a conscious effort to emphasize the second vowel can make the difference between "welder" and "weldor" clear, the precision of enunciation is often not sufficient to clearly indicate to the listener which is which. No such confusion is possible with "welder" and "welding machine". In addition, those who pronounce "welder" differently than "weldor" are not conforming with the English language. The two words are, according to the dictionary, phonetically identical.

The gas tungsten arc welding process was originally used with an inert gas as the arc shielding atmosphere. The term *tungsten inert gas (TIG)* became popular. The later application of non-inert, i.e., active, gases for arc shielding rendered the term *TIG* inaccurate. To remove that discrepancy, the term *tungsten active gas (TAG)* has been proposed by some. With that terminology, the welding of stainless steel with argon is referred to as a "TIG welding process," and if hydrogen is added to the argon shielding gas, the welding process becomes "TAG." If the latter gas mixture is used for welding a noble metal, the welding process would then revert to "TIG." Thus the name of the welding process depends not only on the composition of the shielding gas but also on the base metal composition. Such terminology is no more logical than making the name of the shielded metal arc welding process dependent upon the type of electrode covering and the composition of the base metal.

The proponents of TIG cite its simplicity, brevity, and ease of pronunciation. Tungsten inert gas, by itself, is rather meaningless. Only when the word "welding" is added, is the term complete and may be legitimately compared with gas tungsten arc welding. The term *TIGW* then loses some of its cited advantages.

Arguments similar to those made in support of GTAW, also apply to gas metal arc welding (GMAW) versus metal inert gas welding (MIGW). Both GTAW and GMAW are part of a coherent letter designation system that has been developed by the Definitions Subcommittee for all of the welding and allied processes. Haphazard changes cannot be made without damage to the letter designation system as a whole. That fact is seldom considered by those of the TIG-MIG school.

The Definitions Subcommittee prefers the terms *gas metal arc welding and gas tungsten arc welding*, with modifiers to denote the variations of the processes. In this case, we have made an exception and chosen not to join the reputed majority, in the hope that logic will ultimately prevail.

The origin of the term *diffusion bonding* is unknown, but its widest proliferation may be found in the aircraft and associated industries. Welds sometime fail — of course, unwelded structures also are not immune to failure — and the resulting prejudice has been a contributing cause of welding not reaching its full potential in aircraft construction. It was thought by some that if diffusion welding were given a different name, the aversion to welding would be overcome. It was not, but welding terminology remains plagued by the term *diffusion bonding* and a multitude of corollary terms spawned by the bonding fad.

The Definitions Subcommittee discourages the use of the term "bonding" for "welding". We reserve the term *bonding* for the joining and allied processes, where either an adhesive bond or a mechanical bond is predominant at the interface created by the process actions, i.e., adhesive bonding and thermal spraying. When an atomic bond between the atoms at that interface is predominant, the resulting joint is called a *weld*, and the process that produced that joint is called *welding*, without regard to whether the weld interface is created as a result of fusion or in the solid state. The interatomic bond existing between metal atoms at the weld interface of a fusion weld is no different than that at the weld interface of a solid-state weld.

The term *diffusion welding* is consistent with international custom. The translation of that joining process from any language of the industrial nations into English has for many years been diffusion welding - not diffusion bonding. The recent replacement by the British Standards Institution (*Welding Terms and Symbols*, BS499, Part 1. Glossary for welding, brazing and thermal cutting, 1983) of diffusion bonding by diffusion welding means that *diffusion welding*, rather than "diffusion bonding," is now universally accepted as a part of standard welding terminology.

The Definitions Subcommittee has no illusions about the widespread use of terminology other than the standard welding terminology of A3.0. It would be both futile and arrogant to advocate the eradication of the often picturesque and sometimes fascinating words of those dialects, and our members are neither fools nor supermen. It is only common sense to use the language most appropriate to the occasion; for example, most don't use the same words at home as in the workplace.

Since the primary purpose of language is the communication of concepts, and precise terminology is sometimes cumbersome to use and is hard to understand for those who know only a dialect, when speaking informally or conversationally, it is sometimes better to use less precise terminology if the meaning is clear to the listener. However, a dialect is no longer adequate when one leaves his or her own small circle, unless the dialect of each new circle is learned, and the listener understands just which circle the speaker is in at any particular time.

It is less confusing and more efficient to use a more universal welding terminology, whenever there is any doubt about the listener's specialties. That is where A3.0 comes in, a common welding terminology that makes it possible for anyone to communicate with everybody else in the welding world with only two welding languages; one with the terms of his or her own dialect for local use, and one with the standard terms of A3.0 for a wider audience.

Despite the fact that dialect does have its legitimate and proper use, the intended readership of any publication about welding is seldom so limited that anything less than the standard welding terminology of A3.0 is sufficient.

When terms from A3.0 are included in the glossary of other documents, it is intended that the definitions be identical to those in A3.0, except that the references may be changed if appropriate. A common welding terminology cannot be maintained if one tinkers with what is published in A3.0, to arrive at something that is believed to be a little better for one's own particular limited use. First, the glossary writer changes the definition, then the reviewer makes a few more changes, and the editor may feel like making a few changes also. That definition, which is no longer an A3.0 definition, is then adopted by other glossary writers. As this process continues, the definition departs further and further from the original A3.0 definition, and eventually several different, often conflicting and inaccurate, definitions for the same term result.

If we in the welding industry are to all mean the same thing with a given term, then prose that is consistent with the meaning of terms as they are defined in standard welding terminology, A3.0, must be used, rather than modify the standard definitions to make them fit one's own use of those terms. In many years of reviewing welding documents, it has been found quite practical, with only rare exceptions, to use language that is in conformance with the terms and definitions as they are given in A3.0.

There is such a thing as acceptable tinkering. The Definitions Subcommittee itself is not above that; sometimes it is done in response to comments from outside and sometimes on its own. It is in the nature of things to change, and a living welding terminology must adjust to those changes. In addition, although one of our precepts is that if it works, don't fix it, errors are found in the current A3.0 terminology that must be corrected. There is a vital difference between modifications by the Definitions Committee and those of others. It has little to do with the fact that our assignment is to write definitions, or that our members, through long experience, may be a bit better at it than other welding specialists. It has everything to do with the fact that the modifications proposed by the Definitions Subcommittee are controlled by a lengthy and extensive approval process. Nothing is published in A3.0, until it has been approved by the American Welding Society.

No discussion of terminology would be complete without something on jargon and synonyms.

Jargon, the technical terminology of a special group, is unavoidable in welding language, since that is sometimes the best way to say what is intended, or, it has been in such widespread use for so long that it is impractical to change. With those exceptions, conventional terms should be used; those are understood not only by the welding specialist, but also by all who understand the English language. Furthermore, the "special group" referred to above, with rare exceptions, is the entire welding industry, not a segment. We cannot all speak the same language, if each subgroup or segment has its own independent jargon. There is sufficient jargon in A3.0 to take care of most needs.

Synonyms are not excluded from A3.0; many are already there and more will be added. Where true synonyms, different words that mean the same thing, are concerned, the Definitions Subcommittee does select one of those synonyms as the standard term, while the remainder become nonstandard terms. The principle here is that if the same thing can be said with fewer different words, the understanding of those words is less difficult.

Does the limitation to standard terminology in welding prose stifle originality and freedom of expression? Yes, to some extent. But the Definitions Subcommittee believes that that is a small price to pay for the advantages of a common welding terminology. The welding fraternity is better served by an adherence to standard welding terminology so that we all speak the same language, thereby reducing misinterpretation. There is an essential difference between the needs of welding language and that of nontechnical language. Since the purpose of welding prose is to communicate technical information, there is less need for interesting or persuasive language than in the case of nontechnical literature where the purpose is to entertain, proselytise, or sell. Only a minor portion of an article on welding consists of standard welding terms. There remains available a vast body of English language whose complexities and vagaries are sufficient to test our writing and speaking skills without the addition of uncertainty about the meaning of welding terms. The versatility and richness of that language allows authors ample opportunity to express themselves in interesting and original prose without the savaging of standard welding terminology.

The Definitions Subcommittee does not judge the preferability, acceptability, or correctness of any term. Those determinations are left to reflect the opinion of the welding terminology user. There is one exception. When the use of a nonstandard term may endanger personal safety, that term is defined as both nonstandard and incorrect. The Definitions Subcommittee has neither the authority or the wish to dictate welding terminology, but we do consider it within our province to establish standard terms and nonstandard terms. If the decision is made to use standard welding terminology, there is no logical alternative to A3.0. No other formal, comprehensive body of standard welding terms and definitions is published in this country.

It should not be forgotten that A3.0 is a success. There is increasing acceptance by the American Welding Society and other organizations of that welding terminology. An editorial in the August 1986 *Welding Journal* by the editor lends strong and effective support to the standard welding terminology of A3.0. The Department of Defense long ago abandoned their own welding terminology standard in favor of A3.0. The American Society for Metals has adopted A3.0 for their glossary in Volume 6, Welding, Brazing, and Soldering, of the *Metals Handbook*. The welding terminology of Section IX, Welding and Brazing Qualifications, of the *Boiler and Pressure Vessel Code*, published by the American Society of Mechanical Engineers, is based on A3.0. When any standard is unacceptable to an appreciable segment of its users, an alternate will eventually be developed. For many years, there has been no worthy competition to A3.0. It can only be concluded that most of the welding terminology of A3.0 is acceptable to the majority.

For standard terms to be effective, they should be used consistently at every opportunity. Unilateral decisions to change standard terms and definitions or to use nonstandard terms, more often than not, lead to confusion and disorder.

Those who find what they believe to be an error of either commission or omission in the welding terms and definitions in A3.0 are urged to inform the secretary of the Definitions and Symbols Committee at the American Welding Society of their views. All comments will be objectively evaluated. Changes that would result in significant improvement in standard welding terminology will be made to accommodate proposals. Only through cooperation among those we serve, members of the welding fraternity and the Definitions Subcommittee, is it possible to maintain a common welding terminology, which should be the goal of all who wish to promote better understanding of welding language.

Definitions Subcommittee
Committee on Definitions and Symbols
American Welding Society
Miami, Florida
August 1987

Annex B

A Guide to A3.0

(This Annex is not a part of ANSI/AWS A3.0-94, *Standard Welding Terms and Definitions,* but is included for information purposes only.)

Terms

(1) Any term directly related to welding or allied areas, which has a meaning more specialized or restricted than that given in the standard desk-size dictionary, is defined.

(2) Multiple-word terms whose meaning as related to welding are not clear from a combination of dictionary and A3.0 definitions are defined. Terms for which the combination correctly conveys the meaning for welding usage are not included.

Examples of the former are automatic welding, contact tube, root surface, shielding gas, and workpiece connection. Examples of the latter are arc welder, brazed joint, inert shielding gas, and metal drum welding flux.

(3) Only one form of a term is defined, except as addressed in (14)(a) and (14)(b).

Examples: The term *joint* is defined; other forms such as joints, joining, joined, and join are not. The term *weldability* is defined; other forms such as weldable and unweldable are not.

(4) Terms are categorized as either standard or nonstandard. No other designation, such as preferred or nonpreferred, acceptable or nonacceptable, correct or incorrect is used, except that when the misuse of a term may endanger personal safety, the term is identified as both nonstandard and incorrect.

Example:

ground lead. A nonstandard and incorrect term when used for **workpiece lead**.

(5) A term that has limited and clearly definable applicability includes the area of applicability, in italic type, preceded by a comma, immediately following the term. If either the term or definition reveals the application area, the italicized expression is omitted.

An example of the former:

accelerating potential, *electron beam welding and cutting.* The potential that imparts velocity to the electrons.

Examples of the latter:

arc plasma. A gas that has been heated by an arc to at least a partially ionized condition, enabling it to conduct an electric current.

dynamic electrode force. The force exerted by electrodes on the workpieces during the actual welding cycle in making spot, seam, or projection welds by resistance welding.

(6) No term has more than one definition, except for terms that may be delimited to more than one application.

Example:

horizontal welding position, *fillet weld.* The welding position in which the weld is on the upper side of an approximately horizontal surface and against an approximately vertical surface.

horizontal welding position, *groove weld.* The welding position in which the weld face lies in an approximately vertical plane and the weld axis at the point of welding is approximately horizontal.

(7) Where a verb is commonly used and treated as a noun, the term is stated in the form of a gerund (ending in "ing") and the definition is expressed accordingly. A verb is stated in the infinitive form and identified as such by placing a comma and the letter *v* in italic type after the term. Definitions of verbs begin with the word "to" and are expressed accordingly.

Example:

boxing. The continuation of a fillet weld around a corner of a member as an extension of the principal weld.

(8) A term that is an adjective is identified as such by placing a comma and *adj.* in italic type after the term. The definition is an adjectival phrase, and is not a complete sentence.

Example:

as-welded, *adj.* pertaining to weld metal, welded joints, and weldments after welding, but prior to any subsequent thermal, mechanical, or chemical treatments.

(9) A term that is a noun is stated in the singular form.

Definitions

(10) The essential elements of a term and definition are the term, a period, and one succinct and technically correct sentence to convey the fact or concept represented by the term. The term and basic definition are complete in one sentence when a simple verb such as "is" or "means" is substituted for the period. The definition does not repeat the complete term. A term and basic definition form a genus-species-differentia classical definition whenever possible.

Example:

liquidus. The lowest temperature at which a metal or an alloy is completely liquid.

The term *liquidus* is one species of the genus *temperature*. The remainder of the definition is the differentia that distinguishes this species from all other species, e.g., solidus and preheat temperature, within the temperature genus.

A second example:

soldering iron. A soldering tool having an internally or externally heated metal bit usually made of copper.

This example is comparable to the first, with "soldering iron" being the species and "soldering tool" the genus.

(a) When more appropriate, a definition by extension, which defines a term by enumeration of its parts or of the species for which it is the genus, is used.

Example:

composite electrode. A generic term for multicomponent filler metal electrodes in various physical forms such as stranded wires, tubes, and covered wire.

(11) Supplementary information, in the form of complete sentences, may be included after the basic definition. However, developing this into an encyclopedic discussion is avoided. Unless required for clarity, handbook information and requirements of standards are not included.

Example: (the second sentence is acceptable, the third is not)

Ferrite Number (FN). An arbitrary, standardized value designating the ferrite content of an austenitic stainles steel weld metal. It should be used in place of percent ferrite or volume percent ferrite on a direct replacement basis. See the latest edition of ANSI/AWS A4.2, *Standard Procedures for Calibrating Magnetic Instruments to Measure the Delta Ferrite Content of Austenitic Stainless Steel Weld Metal.*

(12) Only one term is defined in a single location.

(13) Definitions include only defined terms (in either A3.0 or the dictionary), multiple-word partial terms, primary terms, and complete terms; not secondary or single-word partial terms.

(14) All standard terms are completely defined, and the definition does not consist of only a cross reference to another term, except as follows:

(a) Where context makes the meaning of a partial term clear, the partial term is defined by cross referencing the complete term.

Example of a single-word partial term:

cylinder. See **gas cylinder.**

gas cylinder. A portable container used for transportation and storage of compressed gas.

Example of a multiple-word partial term:

welding torch. See **gas tungsten arc welding torch, oxyfuel gas welding torch,** and **plasma arc welding torch.**

gas tungsten arc welding torch. A device used to transfer current to a fixed welding electrode, position the electrode, and direct the flow of shielding gas.

oxyfuel gas welding torch. A device used in oxyfuel gas welding, torch brazing, and torch soldering for directing the heating flame produced by the controlled combustion of fuel gases.

plasma arc welding torch. A device used to transfer current to a fixed welding electrode, position the electrode, and direct the flow of shielding gas and orifice gas.

(b) Where two forms of the same term are in common use and both are acceptable, the secondary form is defined by a cross reference to the primary form, which has a complete definition.

Example:

weld face. The exposed surface of a weld on the side from which welding was done.

face of weld. See **weld face.**

(c) When the meaning of a term is self evident, but a figure is useful, the term is defined by a cross reference to a figure.

Example:

weld metal crack. See Figure 33.

(d) A multiple-word term is stated as it is normally written, accompanied by a definition of the basic term as a cross reference to the multiple-word term.

Examples:

arc welding deposition efficiency. The ratio of the weight of filler metal deposited in the weld metal to the weight of filler metal melted, expressed in percent.

thermal spraying deposition efficiency. The ratio of the weight of thermal spray deposit to the weight of surfacing material sprayed, expressed in percent.

deposition efficiency. See **arc welding deposition efficiency** and **thermal spraying deposition efficiency.**

(e) Some multiple-word terms and their definitions may be made applicable to additional terms and definitions by an obvious substitution of either the basic term or its modifier in both the multiple-word term and its definition. The additional terms are included in the definition of the multiple-word term as variations of that term, accompanied by definitions of the additional terms as cross references to the multiple-word term.

An example of the substitution of the basic term:

automatic welding. Welding with equipment that requires only occasional or no observation of the welding, and no manual adjustment of the equipment controls. Variations of this term are **automatic brazing, automatic soldering, automatic thermal cutting,** and **automatic thermal spraying.** See also **adaptive control welding, manual welding, mechanized welding, robotic welding,** and **semiautomatic welding.**

automatic brazing. See **automatic welding.**

automatic soldering. See **automatic welding.**

automatic thermal cutting. See **automatic welding.**

automatic thermal spraying. See **automatic welding.**

An example of the substitution of the modifier:

thermal cutting operator. One who operates automatic, mechanized, or robotic thermal cutting equipment. Variations of this term are **arc cutting operator, electron beam cutting operator, laser beam cutting operator,** and **oxygen cutting operator.**

arc cutting operator. See **thermal cutting operator.**

electron beam cutting operator. See **thermal cutting operator.**

laser beam cutting operator. See **thermal cutting operator.**

oxygen cutting operator. See **thermal cutting operator.**

(15) Abbreviations are not used in definitions. This includes letter designations of the welding and allied processes.

(16) Units of measurement are not included in definitions.

(17) The definition of a nonstandard term starts with the phrase, "A nonstandard term for," when the term has no use as a standard term, or, with the phrase, "A nonstandard term when used for," when the term is nonstandard for the stated purpose, but is a standard term when used for other purposes. In each case, the introductory phrase is followed by the appropriate standard term or a description of the term use.

Examples using the first phrase:

diffusion bonding. A nonstandard term for **diffusion brazing** and **diffusion welding.**

groove weld throat. A nonstandard term for **groove weld size.**

hydrogen brazing. A nonstandard term for any brazing process that takes place in a hydrogen atmosphere.

Examples using the second phrase:

bottle. A nonstandard term when used for **gas cylinder.**

lead burning. A nonstandard term when used for the welding of lead.

metallizing. A nonstandard term when used for **thermal spraying** or the application of a metal coating.

(18) Nonstandard terms are not used or cross referenced in definitions.

Style and Usage

(19) Terms are arranged in accordance with the dictionary method. That is, the terms are listed alphabetically letter-by-letter, beginning with the first letter of the first word and continuing to the second and subsequent words.

(20) Terms should not be included as groups; however, if an exception is made, each term within the group is also listed in alphabetical order.

(21) Standard terms are printed in boldface. The use of boldface is restricted to standard terms when they are

(a) The term being defined,
(b) Given in a definition cross reference,
(c) The standard term given in the definition of a nonstandard term.

Examples:

weld reinforcement. Weld metal in excess of the quantity required to fill a joint. See also **face reinforcement** and **root reinforcement.**

TIG welding. A nonstandard term for **gas tungsten arc welding.**

furnace brazing (FB). A brazing process in which the workpieces are placed in a furnace and heated to the brazing temperature.

The terms **weld reinforcement** and **furnace brazing** as the first and third terms being defined above are printed in boldface, in accordance with 21(a).

The terms **face reinforcement** and **root reinforcement** in the first definition above are printed in boldface, in accordance with 21(b).

The term **gas tungsten arc welding** in the second definition given above is printed in boldface, in accordance with 21(c).

The terms *weld metal* and *joint* in the first definition given above, the term *TIG welding* as the second term being defined above, and the terms *brazing, workpieces*, and *brazing temperature* in the third definition given above are printed in lightface, since those terms are not included in 21 as uses for boldface.

(22) The word "deposit," or any of its derivatives, is used only in connection with the terms filler metal or surfacing metal. Use with such terms as *weld metal, weld bead, weld*, etc. is nonstandard.

(23) All cross references to figures and tables begin with the word, "See." All cross references to terms begin with the words, "See also," except as explained in 14. Cross references are stated in the order of figures, tables, and terms, which are stated in alphabetical order.

(24) Definitions of terms describing weld conditions shall limit the use of wording indicating acceptability or rejectability of those conditions to those words necessary to accurately define those terms.